"Professor Sternglass's courageous voice has helped keep alive the debate on the health effects of low-level radiation, a debate that the military, the nuclear industry, and even some biologists and physicians have tried to bury. His new book, *Secret Fallout: Low-Level Radiation from Hiroshima to Three Mile Island,* is an important new contribution to that debate and should be required reading for all who are concerned with their own health, the health of their children, and of their children's children."

> Victor W. Sidel, M.D.
> Professor and Chairperson
> Department of Social Medicine
> Montefiore Hospital and Medical Center
> Albert Einstein College of Medicine

Dr. Ernest Sternglass is Professor of Radiology, specializing in radiological physics, at the University of Pittsburgh Medical School, as well as Adjunct Professor in the Department of History and Philosophy of Science at Indiana University, Bloomington. He is past president of the Pittsburgh Chapter of the Federation of American Scientists, a Fellow of the American Physical Society, a member of the Radiological Society of North America and the American Association of Physicists in Medicine. He has testified on low-level radiation before the Joint Committee on Atomic Energy and many other groups both here and abroad.

Dr. George Wald, Nobel Laureate in Physiology and Medicine, is Professor of Biology at Harvard.

Ernest J. Sternglass

SECRET FALLOUT

LOW-LEVEL RADIATION FROM HIROSHIMA TO THREE MILE ISLAND

McGraw-Hill Book Company

New York   St. Louis   San Francisco
Auckland   Bogotá   Guatemala   Hamburg
Johannesburg   Lisbon   London   Madrid   Mexico
Montreal   New Delhi   Panama
Paris   San Juan   São Paulo   Singapore   Sydney   Tokyo   Toronto

Copyright © 1972, 1981 by Ernest J. Sternglass

All rights reserved. Printed in the United States of America. No part of this publication may be reproduced, stored in a retrieval system, or transmitted, in any form or by any means, electronic, mechanical, photocopying, recording, or otherwise, without the prior written permission of the publisher.

This book is an expanded verison of *Low-Level Radiation,* first published in 1972 by Ballantine Books.

First McGraw-Hill Paperback edition, 1981

1 2 3 4 5 6 7 8 9 0   FGFG   8 6 5 4 3 2 1

LIBRARY OF CONGRESS CATALOGING IN PUBLICATION DATA

Sternglass, Ernest J
Secret fallout.
Ed. of 1972 published under title: Low-level
radiation.
Bibliography: p.
Includes index.
1.   Radioactive pollution—Toxicology.   2.   Radiation—Toxicology.   I.   Title.   II.   Title:
Low-level radiation from Hiroshima to Three Mile
Island.
RA569.S69 1981   616.9'897   80–22390
ISBN 0–07–061242–0

Book design by Roberta Rezk

The author acknowledges the following sources for permission:

The Washington Post for Bill Curry's "A-Test Officials Feared Outcry after Health Study."
The New York Times for James Reston's "The Present Danger."
© 1978 by The New York Times Company. Reprinted by permission.
Harrowsmith Magazine for Tom Pawlick's "The Silent Toll."
Beaver County Times for Joel Griffiths' "State Panel Questions, Radiation, Safety."

The duty to endure
gives us the right to know.
—Jean Rostand

DISCARDED
PASADENA CITY COLLEGE

For Marilyn,
and our children
Daniel and Susan

Contents

Acknowledgments

THIS BOOK could not have come into being without the understanding, concern and support of many individuals whose help I was privileged to receive during the years in which the events recounted here took place.

Although it is impossible to list all those to whom I have become deeply indebted, there are a few individuals whose help went far beyond anything I shall ever be able to acknowledge adequately.

First among these is my wife Marilyn, who not only stood by my side throughout these years, but also provided the constant counsel, encouragement and understanding needed in the long and arduous task of writing this book.

And it was the great personal dedication of my editor, Joel Griffiths, that shaped the first edition of the book, helping immeasurably in the difficult task of explaining for the non-scientist the complex scientific and technical arguments underlying the events described. Thus, though the responsibility for the accuracy of the facts and their interpretation must remain mine, whatever success this book may have in clarifying the nature of the scientific problem and the dangers arising from the misuse of nuclear radiation will be to a large extent a reflection of his efforts.

Among the many others who contributed importantly to bringing this book into being, I must express my indebtedness to Larry Bogart, long-time conservationist and founder of the National League to Stop Environmental Pollution, who together with Leo Goodman of the United Automobile Workers was responsible for first drawing my attention to the full hazard of an unchecked nuclear technology.

In these researches, I was greatly aided by two of my colleagues, Dr. Donald Sashin and Ronald Rocchio, who worked out the computer programs that made the analysis of the vast amount of statistical data possible, as well as by Michael Szulman, Diane Gaye, Mitchel Margolis and Debbie Conant, without whose dedication in patiently collecting and analyzing the data the task would have been insuperable.

In the collection of the basic data, I was also generously helped by a number of young volunteer student assistants who spent long hours in the library. My gratitude goes to all of them, and in particular to David and Harold Colker, Randolph Strothman, and Gary Harris, whose important contributions were particularly appreciated.

Last but not least, I must express my deep appreciation to my secretary, Judy Czachowski, who patiently suffered through the agonies of gathering the data, preparing papers and repeated retyping of the original manuscript, and without whose cheerful help and dedication the task would have been vastly more difficult.

The present, greatly expanded edition owes its existence to my new editor, Joanne Dolinar, who persuaded me that it was important to tell the story of the developments in the ten years since the book's original publication. For her persistence in this I am deeply grateful. I am also grateful to my secretary, Nancy Siegel, who greatly eased the task of completing the expanded version of the book with her unwavering patience in tackling the never-ending pages with their illegible revisions.

Preface

WHEN I UNDERTOOK to write the first edition of this book, originally published in 1972 under the title *Low-Level Radiation,* my primary concern was with the health effects of worldwide fallout from nuclear weapons, particularly on the developing infant in the mother's womb.

At that time I also discussed the first evidence for possible health effects of routine releases of radioactivity from nuclear reactors in their ordinary day-to-day operation.

In the ten years that have intervened since then, my concerns about the safety of nuclear plants have unfortunately been reinforced far more than I could have anticipated. Not only in the accident at Three Mile Island, whose likely effects on human health are discussed in the present book, but also in the normal operations of many other nuclear plants, there is now growing evidence for rising infant mortality and damage to the newborn. In the decade that has passed, cancer rates increased most sharply in areas closest to the nuclear reactors whose radioactive gas releases were found to rise most strongly, following the earlier pattern of death rates among the newborn described in the original book.

The first fourteen chapters have been left nearly unchanged, while the rest of the present book brings the story up to the present time. It deals with the newly disclosed evidence that the possibility of serious health damage from weapons testing was long known to our government. It also presents the evidence for widespread damage to the learning abilities of the children born in areas of heavy fallout during the period of massive nuclear weapons testing.

What emerges is that in order for major governments to be able to continue threatening the use of their ever-growing stockpiles of weapons to fight and win nuclear wars rather than merely to deter them, they must keep from their own people the severity of the biological damage already done to their children by past nuclear testing and the releases from nuclear reactors near their homes.

It is to focus attention on the need to end this hidden threat to the future of human life on this globe that this new edition has been prepared.

Ernest J. Sternglass
Pittsburgh
July 1980

Introduction

I SHOULD LIKE TO START with a few words concerning the human condition, and go on with a little of the special problems of the author in writing this book.

As a scientist I take a long view of history: 20 billion years of this universe; 6 billion years of the solar system; 4.7 billion years of the planet Earth; 3 billion years of life on Earth; something like 3 million years of something like human life; 10,000 years of civilization; and then—something happened.

In 1976 we celebrated the bicentennial of American independence. That independence was an interesting event, but not nearly as important even to Americans as something else that was happening at the same time. That was the Industrial Revolution. At first it promised humanity endless leisure and abundance. But a half-century ago it turned life-threatening on the grand scale; and now killing and destruction are the biggest business in the world. Military expenditures worldwide in 1979 were over $460 billion, and rising rapidly. The simple reality is that a trivial two hundred years of the Industrial Revolution have brought the human species to the brink of self-extinction.

Nuclear war is the most immediate threat. Just the "strategic" nuclear weapons—the big ones, in the megaton range*—now stockpiled by the U.S. and the Soviet Union add up to the explosive equivalent of about 16 million tons of TNT. There are just over four billion persons on the Earth, so about 4 tons of TNT for every man, woman and child in the world. In addition each superpower has stored tens of thousands of so-called "tactical" nuclear weapons, and the material to make hundreds of thousands more.

So I had better say what a tactical nuclear weapon is. The bomb that in a moment leveled the city of Hiroshima and by the end of that year—1945—had killed 140,000 persons rates in the present arsenals as a pitifully small "tactical" weapon, a mere 12.5 kilotons. For comparison the Titan missile whose fuel recently blew up in its silo in Arkansas had perhaps 100 times that explosive power.

But the explosive power—the blast and heat and radiation—are just the immediate release of nuclear weapons. There is also the mushroom cloud of radioactive fallout that enters the atmosphere and stratosphere and eventually covers the entire globe. This goes on showering the Earth with potentially lethal ionizing radiation, and every rain and snowfall brings down radioactive elements to be inhaled, and by entering the food chain, ingested. And that goes on and on, from the comparatively short-lived iodine-131 and strontium-89, dangerous for 6 months to a year, to plutonium-239, perhaps the most toxic substance known, whose half-life—the time it takes for its radioactivity to half-decay—is 24,400 years. That remains dangerous, in human terms, forever.

* Nuclear weapons are graded in terms of the equivalent explosive power in tons of TNT: kilotons, which are thousands of tons; or megatons, millions of tons.

An interesting dialogue taking place in the Atomic Scientists Bulletin raises the question: would anyone survive? If the present stockpiles of nuclear weapons were used, would any human beings be left on the Earth? That is at least questionable, and we would take a lot of the rest of life on this planet with us.

Directly out of the business of nuclear weapons came the business of nuclear power, heralded in our country with the slogan, *Atoms for Peace.* Even that innocent-sounding slogan is part of the endless pattern of public deception that surrounds the entire nuclear enterprise. Let me interject a present example that poses the relationship nicely. In our country the entire hydrogen bomb enterprise—both R and D and production—is not under the Department of Defense, but the Department of Energy. It goes, not into the Defense budget, but the Energy budget. It is by far the largest item in that budget, consuming well over one-third of it. The next largest item in it is nuclear power.

Nuclear power and nuclear weapons are two sides of the same coin. Nuclear power is life-threatening in three independent ways, each in itself formidable.

First is the threat of accident in nuclear power plants. This book tells in some detail the story of the accident at Three Mile Island. But one didn't have to wait for that to know that nuclear power plants— unlike what the public has been told—are thoroughly accident-prone. Those great realists, the American insurance companies, refused from the beginning to insure nuclear power plants. Hence we have the Price-Anderson Act, renewed by Congress every 10 years since 1957, which lays the bulk of the liability in the event of nuclear accident on "the government"—i.e., on the taxpayers.

The second life-threatening property is that every nuclear reactor now in operation produces the artificial element plutonium-239 as by-product. This is not only, as already said, perhaps the most toxic substance known. It is also the most convenient material from which to make fission bombs. The "trigger quantity"—the smallest amount from which one can make a workable atom bomb—is 2 kilograms, 4 2/5 pounds. You could carry that, and safely, in a grocery bag. To make a Hiroshima-size bomb would take 6–7 kilograms, say about 14 pounds. You'd need a shopping bag for that. Every nation that now possesses a nuclear reactor can, if it chooses, begin to make nuclear weapons. It is expected that within the coming decade perhaps a dozen more nations than now possess them will exercise this option. It should be added that plutonium provides the trigger at the core of all hydrogen bombs, and in some also the shell.

The third life-threatening aspect of both nuclear power and weapons involves the disposal of nuclear wastes. No one knows what to do with them. The periodic meetings of international experts have so far yielded no credible solution.

In my opinion the entire nuclear enterprise, both power and weapons, represents a wrong turn for humanity, a development that cannot be tamed, that remains life-threatening not only in all its present manifestations, but all future developments that have been contemplated.

Meanwhile the public is subjected to a continuous barrage of propaganda and misinformation designed to reconcile it to an increasingly problematical and expensive support of both nuclear power and weapons. The weapons, ostensibly for our security, are of course the principle source of our insecurity; and the nuclear power, that we are told we need for energy, supplies in 1980 only about 12% of our consumption of electricity, hence only about 2% of our total energy consumption, at a still unreckonable cost in both health and money.

The author of such a book as this is under constant attack, not only from the expected sources in industry and government, but from certain quarters in the science establishment. I have heard at times from fellow scientists, some indeed on the same side as Professor Sternglass in opposing the spread of ionizing radiations, the somewhat querulous comment, "I don't like his statistics." That would impress me more if I had ever met anyone who liked anyone else's statistics. That's the way with statistics: they are highly individual. Sternglass has an exuberant way with them. At times in this book I had the feeling he was going a little far. But then I never could be sure, once I had read over carefully what he was saying, that it was *too* far. The truth is that once one starts down this path, it's hard to know where or whether to stop. And on the fundamental issues, Sternglass is dealing with a very strong case. I think that it is by now beyond doubt that ionizing radiations at all levels involve serious risks to health, causing increased chances of cancers, leukemia and genetic effects. There is no threshold: a little, however little, causes some increased risk, and more causes more risk. There is no level that fails to be potentially harmful. From that point of view the existence of an official so-called "permissible level" is misleading. A "permissible level" of radiation only has meaning in cost benefit accounting; and that would mean more if the costs and benefits involved the same parties. Unfortunately they usually do not: one group—workers, general public—commonly bear the costs; and another, quite different group—ownership, management, government—shares the benefits. Having to deal with a lot of official talk about "permissible levels" of radiation at the time of Three Mile Island, I took to saying, "Every dose is an overdose." I believe that to be true as a statement, not necessarily of overt effect, but of risk.

GEORGE WALD
Paris, October 28, 1980

1

Thunderstorm
in Troy

ON MONDAY MORNING, April 27, 1953, the small group of
students in Professor Herbert Clark's radiochemistry class at
Rensselaer Polytechnic Institute walked into the metal shack
that served as their laboratory, located high on a hill overlook-
ing the city of Troy in upper New York State. The students
set about making preparations for the day's experiments, but
then Professor Clark interrupted to draw their attention to
something unusual. All the Geiger counters were registering
radiation at many times the natural rate.

Since instruments nearest the outer walls were giving the
highest readings, several students immediately went outside
with a portable Geiger counter. At once they found that wherever they walked, the count rate on the ground was far above
normal, in some places a thousand times as high. In particular,
beneath the spout of the gutters that carried the rainwater
down off the roof of the shack, the needle gave a disconcertingly
high reading. Evidently the previous night's heavy rains had
brought down large amounts of radioactivity.

Dr. Clark quickly guessed the source. Such high readings
could only have come from heavy deposits of fallout, the drifting

clouds of radioactive debris created by the explosion of a nuclear bomb in the atmosphere. To verify his guess, he phoned John Harley, a friend and former colleague who now worked for the U.S. Atomic Energy Commission's Health and Safety Laboratory in New York City. As one of Dr. Clark's students recalled the story many years later, Harley's first reaction was that Clark must be kidding, and, expressing amused disbelief, he hung up. But a few minutes later, New York called back. Dr. Clark summarized the details of the morning's measurements: how the count rate from the gamma radiation on the ground was anywhere from ten to five hundred times normal, how the activity from beta rays had gone up even more, and how "hot spots" beneath rainspouts and in puddles on the pavement showed still higher readings, much higher than he had ever observed after other nuclear tests, when it had been hard to measure any additional radioactivity at all. Thoroughly alarmed, the director of the New York Laboratory, Dr. Merrill Eisenbud, promised to check personally into the situation, to send some of his top people to make their own measurements on the spot, and to take any steps that might be called for to protect the public health.

For, as Dr. Clark had just learned, there had indeed been a recent atomic bomb test, conducted by the AEC in Nevada two days earlier. The bomb, code-named Simon and equivalent in power to 43,000 tons of TNT, had been detonated in the atmosphere some 300 feet above the desert. The upper portion of the mushroom cloud had reached an altitude of about 30,000 or 40,000 feet and then drifted 2300 miles across the United States in a northeasterly direction, passing high over Utah, Colorado, Kansas, Missouri, Illinois, Indiana, Ohio, and Pennsylvania before it encountered a severe thunderstorm in progress over most of upstate New York, southern Vermont, and parts of Massachusetts.

The storm was an extraordinarily violent one, accompanied by extremely high winds, hail, and torrential rains that flooded streets and basements, undermined foundations, and caused heavy damage to trees and houses. It was one of the heaviest flash storms Dr. Clark could remember. The sudden cloudburst, he surmised, had probably brought much of the fallout down in concentrated form. Dr. Clark quickly put his students to work in an effort to determine just how serious and widespread the danger might be.

Students set out with portable radiation detectors and began measuring the radioactivity on the pavement, on pieces of cloth, on asphalt roof shingles, on burdock leaves and other vegetation—any place it would be likely to collect and adhere. Samples were also taken of water from reservoirs and household taps. Within a matter of hours the students were reporting back from such nearby towns and cities as Watervliet, Mechanicville, Saratoga Springs, Albany, and Schenectady that everywhere the radiation levels were about the same as on the campus. Typical readings were twenty to a hundred times normal, with hot spots up to ten times higher than that.

Now knowing the radiation levels as well as the source and age of the fallout, Dr. Clark could calculate that during the next ten weeks the total gamma radiation dose to the population from the radioactivity in the environment would be, on the average, roughly equivalent to that received from a typical diagnostic X-ray exposure. This was reassuring, since such a dose was not very different from what most people in the world receive each year from the naturally occurring cosmic rays that penetrate the earth's atmosphere. And it was well below the maximum permissible dose limits set by government agencies.

However, there was also the high radioactivity in the rainwater, which was certain to contaminate the reservoirs and thus the tap water. The samples of rainwater collected from a puddle on the campus had shown a radioactivity level of 270,000 micromicrocuries per liter, thousands of times higher than the maximum levels then permitted by AEC standards, which were set at 100 micromicrocuries per liter. Normal drinking water usually had an activity of about 1 micromicrocurie per liter.

There was, accordingly, much apprehension among the students until the samples of actual drinking water from the taps and reservoirs could be analyzed early the next day. When this was done, the first of the tap water samples, taken Monday night, showed an activity of 2630 micromicrocuries per liter— not as great as was feared, yet still well in excess of the limit. But by that evening, the same tap gave a sample with a greatly decreased activity of 1210 per liter, while samples from nearby Tomhannock Reservoir ranged from 580 to 960. The radioactive rain was evidently becoming heavily diluted in the reservoir before reaching the taps in the households of Troy.

Thus, all concerned were greatly relieved that the total radiation doses received by the populace would probably turn out to be relatively small. It would not be necessary to filter the drinking water or decontaminate the streets and rooftops by means of elaborate and costly scrubbing procedures, a monumental task in view of the tenacity with which the radioactivity had been found to cling to rough surfaces such as pavement, asphalt shingles, and burdock leaves, and especially to porous materials like paper and cloth. Dr. Clark and his students found that even treatment with hot, concentrated hydrochloric acid— an extreme method—was only partially effective in removing the radioactivity from the objects to which it clung. The class also conducted tests to determine the strength of this radioactivity. Surprisingly, they found that it was comparable to that reported the previous year by the AEC's New York Laboratory for fallout in desert areas only 200 to 500 miles from the point of detonation at the Nevada test site itself.

But the possible health effects of any internal doses that might result from eating, drinking, or breathing the radioactivity were considered negligible by the New York State Health Department and the AEC. And so it was decided that nothing further need be done. An editorial in the local newspaper expressed some concern, but soon the whole incident was forgotten.

Meanwhile, however, Dr. Clark, under contract to the AEC, continued to monitor the levels of radioactivity in the reservoirs, while AEC physicists, using an extremely sensitive gamma-ray detector mounted in an airplane, conducted extensive surveys of the entire region. Detailed reports on the findings were written by the staff of the New York Lab, but, since they were classified "secret," the public never learned of their contents. All that appeared was the following brief statement in the 14th Semi-Annual Report of the Atomic Energy Commission for the first half of 1953:

> After one detonation, unusually heavy fallout was noted as far from Nevada as the Troy-Albany area in New York. Following a heavy rain in that area on the second day after the detonation, the concentration of radioactivity was from 100 to 200 curies per square mile. It is estimated that this level of radioactivity would result in about 0.1 roentgen exposure for the first 13 weeks following the fallout. The exposure has no significance in relation to health.

One fact the AEC did not announce, and the general public did not learn, since it was later published by Dr. Clark in the obscure, highly specialized *Journal of the American Water Works Association,* was that, as the AEC continued its nuclear testing in Nevada during the spring of 1953, further rainouts repeatedly raised the radioactivity in the reservoirs serving Troy to levels comparable to those measured by Dr. Clark and his students the morning after the "Simon" rainout in April.

2

The Unheeded Warning

THE TROY INCIDENT was easily forgotten because, at the time, little was known about the effects of low-level radiation—either from fallout or from other sources. The subject had hardly even been thought about. Scientists generally assumed that such levels were harmless, since they produced no immediately observable effects. During the next few years, however, tremendously improved radiation measurement techniques coupled with detailed laboratory studies revealed many previously unsuspected hazards from fallout. And with these discoveries, the forgotten incident in upstate New York re-emerged and took on great significance.

By 1953, it was already known that many of the radioactive elements (called isotopes) created by an atomic explosion, once they entered the atmosphere in the form of tiny fallout particles, would contaminate food, water, and air and thus find their way into the human body. What was not widely known, however, was the extent to which these isotopes became concentrated in various body organs. Inside the body, they behaved just like their nonradioactive natural counterparts. The isotope strontium, for instance, which is similar to calcium, settled in bones and teeth. Radioactive iodine behaved like regular iodine, seeking out and concentrating in the thyroid gland, an organ which is vital in regulating the growth and functioning of the human body.

It was in the case of iodine that some of the most alarming discoveries were made. In the early 1950s researchers found that iodine became concentrated in the milk of cows that grazed on pasture contaminated with fallout. When people drank the milk, the iodine began building up rapidly in their thyroid glands. Since the thyroid gland is small in size, the concentration was very heavy. Measurements revealed that in any given situation the radiation dose to the adult thyroid would be as much as a hundred times the external dose from the fallout in the outside environment. But far more important were the results of extensive studies conducted at the University of Michigan and published in 1960. These showed that the radiation dose to the thyroids of unborn children and infants was ten to one hundred times higher than that to the adult because of the greater concentration in the smaller thyroids. This discovery held serious implications for the health of the children of Troy. It meant that the doses to their thyroids might have been as much as a hundred to a thousand times higher than those estimated by Dr. Clark and the AEC scientists, who had only considered the overall dose from the fallout in the external environment.

However, by the time these discoveries became widely known, a voluntary halt in atmospheric testing had been agreed upon by the Soviet Union, the United States, and Great Britain, and there was considerable hope that incidents of heavy fallout would never occur again. Thus it seemed less urgent to pursue investigations into the problem. But in 1961, during the Berlin crisis, Russia's detonation of a 100-megaton hydrogen bomb high over Siberia marked the resumption of large-scale atmospheric testing by the nuclear powers, and the levels of radioactivity in air and water once again rose sharply throughout the world. In the weeks that followed, an enormous peak of radioactive iodine was detected in milk throughout the northern hemisphere. As the testing continued, many scientists began to feel it was imperative to find a conclusive answer to the question: Just exactly how harmful was low-level radiation from fallout?

It was in this context that the well-known nuclear physicist Ralph Lapp wrote an article for *Science* magazine in 1962 which first focused attention on the significance of the Albany-Troy incident. Lapp's article showed that radiation doses far larger than those permitted by federal safety guidelines must have been received by the children of Troy and numerous other cities that had been subjected to similar "rainouts" in the early years of testing. The purpose of the article was to point out that the Troy incident provided an excellent opportunity to find out just what the effects of fallout were. The surrounding area's population of half a million persons was large enough to insure that any increase in the normally low incidence of such radiation-caused diseases as thyroid cancer or childhood leukemia would show up. (The normal incidence of leukemia among children under ten years old was about two to three cases per year per 100,000 children. Thus, if any area with only a few thousand children were studied, no cases at all might be found in some years, even if the radiation were strong enough to double the normal expected number.) And the detailed radiation measurements taken by Dr. Clark's students and the AEC meant that relatively accurate estimates could be made of the doses involved.

The study that Lapp proposed had enormous potential ramifications. At the time, many people in government, military, and scientific circles still believed that mankind could survive the levels of fallout that would result from a nuclear war, levels thousands of times greater than those from peacetime testing. The United States had embarked on an extensive civil-defense program based on this belief. But if it were shown that peacetime fallout levels led to a significant increase in fatal diseases, then by implication, nuclear war would probably mean the end of mankind, and thus the vast nuclear war machinery developed by the United States and the Soviet Union would become useless. In the second place, if it were shown that large numbers of children had already died from the effects of fallout, then tremendous public revulsion would probably be generated against *all* activities that released more radioactivity into the environment. These would include not just the testing of nuclear weapons, but also the monumental program planned by many governments and industries throughout the world for the peacetime uses of atomic energy. For nuclear power reactors, atomic gas-mining explosions, and other forms of nuclear engineering all normally release low levels of radioactivity and, in the event of an accident, they entail the risk of much worse. And, finally, those individuals who had been in positions of responsibility would have a terrible guilt to bear for the damage already done.

The appearance of Lapp's article also served to highlight another extraordinary fact. It was then seventeen years since the first atomic explosion at Hiroshima in 1945, yet no large-scale cancer studies such as he proposed had ever been carried out, even though the AEC had long been in possession of detailed fallout data for many areas of the U.S. A great deal of information existed on the effects of high doses of radiation, such as those received by the survivors of the explosions at Hiroshima and Nagasaki, but there was no real evidence regarding low-level effects, either from laboratory animal studies or from direct observations of large human populations. The lack of animal studies was somewhat understandable, since no such experiments could be carried out at the extremely low doses produced by fallout without requiring hundreds of thousands or even millions of animals and many years to detect the small increases of a rare disease such as leukemia. But in the case of humans, such a large study population had already been created by the fallout from years of atomic testing. Yet the AEC had ignored this opportunity to resolve such an important issue. Thus, those who wished to minimize the danger of continued atomic testing could argue, in the absence of data to the contrary, that long-term, low-level exposure such as that from fallout had not been proven to increase fatal diseases.

The absence of such studies was all the more striking because there were already strong indications that such danger existed. It was toward the end of 1955 that Dr. Alice Stewart, head of the Department of Preventive Medicine at Oxford University, first became aware of a sharp rise in leukemia among young children in England. A young statistician in her department, David Hewitt, had discovered that the number of children dying of this cancer of the blood had risen over 50 percent in only a few years. In the United States an increase about twice as large had occurred. One aspect of this rise was extremely puzzling: The leukemia seemed to strike mostly children over two to three years of age—there was little or no increase for younger children. This had not been the situation prior to World War II, when both groups had shown a parallel, much more gradual rise. The question was: What new postwar development could be responsible for the increase in deaths among the older children?

Dr. Stewart undertook a study to find out. With the assistance of health officers throughout England and Wales, she obtained detailed interviews with the mothers of all of the 1694 children in those countries who had died of cancer in the years 1953 to 1955, as well as with an equal number of mothers of healthy children. By May 1957, the analysis of 1299 cases, half of which involved leukemia and the rest mainly brain and kidney tumors, had been completed. The data showed that babies born of mothers who had a series of X-rays of the pelvic region during pregnancy were nearly twice as likely to develop leukemia or another form of cancer, as those born of mothers who had not been X-rayed. As Dr. Stewart noted, the chance of finding such a two-to-one ratio purely as a result of statistical accident was in this case less than one in ten million. Thus, in the paper she published in June 1958, Dr. Stewart concluded that the dose from diagnostic X-rays could produce a clearly detectable increase in childhood cancer when given during pregnancy.

This was an extremely low dose. It was roughly comparable to the dose that most people receive in only a few years from natural background radiation. (Mankind has always lived with a "natural background" of radiation, produced by cosmic rays and various naturally occurring radioactive substances. The annual dose from the radiation averages about 100 millirads.) But still more significant, this dose was comparable with what the pregnant mothers of Albany-Troy must have received from the fallout of the "Simon" test in 1953.

In this connection, there was another finding of Dr. Stewart's study that was even more disturbing. This concerned the timing of the X-rays. Children whose mothers were X-rayed during the first third of their pregnancy were found to be some ten times more likely to develop cancer than those whose mothers were X-rayed toward the end of pregnancy. In other words, the earlier the worse. This finding had much more serious implications for fallout than for medical X-rays. Almost 90 percent of pelvic X-ray examinations occur shortly before delivery time, but since fallout comes down indiscriminately on whole populations, it irradiates unborn children at all stages of development, including the earliest. The fallout hazard was further compounded by the tendency of various radioactive elements, such as iodine and strontium, to concentrate in vital body organs. This meant that the doses to the thyroids and bone marrows of unborn children from fallout could be many times higher than the doses received from diagnostic X-rays by the children in Dr. Stewart's study, which had already nearly doubled the cancer incidence.

But in order to establish a clear cause-and-effect relationship between the X-rays and the additional cancer deaths, there had to be a direct relationship between the amount of radiation received by the fetus and the chance that the child would develop cancer a few years later. And indeed, when Dr. Stewart and David Hewitt examined the available records for the number of X-ray films taken, they found that there were distinctly fewer cancer cases among the children whose mothers had only one X-ray than among those who had four or more. The number of cases where this information was available was too small to establish a conclusive connection between dose and cancer risk, but there was other evidence that supported this general trend. For example, whenever the X-rays had been taken only of other parts of the body, such as the arms and legs, so that only a small quantity of scattered radiation reached the unborn child in the womb, the increase in cancer risk was only about one-fifth as great as in those cases where the abdominal region itself was X-rayed.

These latter observations were in direct contradiction to a belief that was essential to the continuation of all programs for nuclear testing and the peaceful uses of the atom—namely, the so-called "threshold" theory. This theory held that there was a certain low level of radiation exposure, a "threshold," below which no damage would be caused. If this threshold was about the same as the yearly dose from background radiation or from exposure to typical diagnostic X-rays, as various supporters of nuclear programs maintained it was, then there would theoretically be no ill effects from past or present weapons tests, from the radioactive releases of nuclear reactors, or even from the radiation persisting after a nuclear war, since this radiation would probably not exceed the threshold if it were averaged out over a lifetime. But Dr. Stewart's study implied that if there were any safe threshold for unborn children and infants it would have to be less than the dose from a single X-ray picture. And her finding that the risk of cancer seemed to be directly related to the size of the dose suggested that there might not be any safe threshold at all, and that *any* increase in radiation exposure might produce a corresponding increase in the risk. Even if the risk for a certain tiny amount of radiation was extremely small, say, one chance in ten thousand, then if millions of people were exposed to this radiation, hundreds would be likely to get cancer. Fallout had already exposed millions of people to doses comparable to those received by the children in Dr. Stewart's study, and the proliferation of nuclear explosions for peaceful purposes would make this exposure even more extensive.

There was widespread refusal to accept the implications of Dr. Stewart's work. Her findings were regarded as doubtful for such reasons as their dependence on the memories of the mothers as to the number of X-ray exposures received. Other studies were cited that showed no effects from X-rays. It was said that her study was inapplicable to fallout because it had been shown that a specified dose of radiation given all at once— as is the case with a diagnostic X-ray—is more damaging than the same total dose given gradually over a period of weeks, months, or years—as is the case with fallout.

This argument opened up another important area of dis-

agreement about radiation dangers. Were the cancer-causing effects of radiation cumulative? Or did body cells recover? There
was no question that body cells did repair themselves in the
case of such damage as radiation burns, which healed with
the passage of time. Supporters of the threshold theory hypothesized that this would also hold true for cancer. This was another bulwark of the "threshold" theory, for, if such recovery
did take place, then there would indeed exist a level of radiation
low enough so that the body's repair mechanisms could keep
pace with the damage.

However, evidence was soon forthcoming that would refute
the criticisms of Dr. Stewart's study and thereby cast further
doubt on the validity of the threshold theory. After the publication of Dr. Stewart's results, Dr. Brian MacMahon of the School
of Public Health at Harvard University undertook another
study of the relationship between diagnostic X-rays and childhood cancer. He constructed this study so that there would
be no question as to the number of X-rays given to the mothers.
Using the carefully maintained hospital records of 700,000
mothers who delivered their babies in a series of large hospitals
in the northeastern United States between 1947 and 1954, he
compared the risk of cancer for the children of the 70,000
mothers who had received one or more X-rays with the risk
for the children of the remaining 630,000 mothers who had
received no X-rays during pregnancy.

The results of his study, published in 1962, fully confirmed
the findings of Dr. Stewart: There was a clear and highly significant increase in the risk of cancer for the children who had
been X-rayed before birth, and, most important, the risk did
indeed increase with the number of X-rays taken. The overall
risk was somewhat smaller than had been found for the British
children by Dr. Stewart, but this could easily be explained
by the fact that the dose to the mothers in MacMahon's study
from each X-ray picture was substantially lower than for those
in Dr. Stewart's, due to improvements in X-ray technology.
As for the studies cited by critics which did not show any
increase in cancer risk from prenatal X-rays, it developed that
these were all based on small study populations, and even then

the indications were that if these results were carried out to larger numbers they would confirm Stewart and MacMahon.

But there was still one major question that remained unanswered. To what degree were the effects of diagnostic X-rays comparable with those of fallout?

There were already many indications that the effects might be similar. Among these was the fact that had prompted Dr. Stewart to undertake her study in the first place, namely, the evidence that in both the United States and England cancer and leukemia among school-age children had increased sharply beginning a few years after World War II. This was the period when nuclear fallout was first introduced into the atmosphere. And now, Dr. Stewart's and Dr. MacMahon's studies had served to point up the following significant aspects of this increase:

First, the effects of X-rays, although very real, were not strong enough to have caused all of the very large general increase in childhood cancer, which ranged from 50 to 100 percent. Dr. Stewart herself estimated that X-rays could only have accounted for perhaps 5 percent of this increase.

Second, this general increase had taken place only among children older than two or three—exactly the age group that had suffered the greatest effects from X-rays. This suggested that some other form of radiation might be causing the unexplained portion of the increase, since the characteristic age at death was the same.

Third, other possible factors such as the introduction of new drugs, pesticides, or food additives had been ruled out because these factors had been found to be essentially the same for the healthy and afflicted children alike.

But the main reason why it seemed that fallout was at least as effective as X-rays in producing childhood cancer was the growing evidence for a direct relationship between the number of X-ray pictures taken and the risk of cancer. For if the risk increased with each additional picture, as the studies of Stewart and MacMahon indicated it did, then this clearly implied that there was no significant healing of the damage and thus that the cancer-causing effects of radiation were cumulative. This would mean that the effects of a dose received over a period of time from fallout would be similar to those from an equal dose received all at once from X-rays.

Such a direct connection between the amount of radiation absorbed and the likelihood of cancer could be predicted on the basis of a theory developed by Dr. E. B. Lewis of the California Institute of Technology. According to Dr. Lewis, cancer could be triggered if one particle of radiation scored a single bulletlike hit on a crucial DNA molecule in the chromosomes of a cell. The DNA contains the genetic code that controls the functioning and reproduction of the cell. If it were damaged by a particle of radiation, this might disrupt the governing mechanism and cause the cell to begin the unlimited growth which characterizes cancer.

The significance of this theory was twofold. First, it was already established that such damage to the DNA was one of the ways that radiation produced hereditary or genetic damage in the female ova and male sperm cells—the type of damage that results in malformations and other harmful mutations in offspring. What Dr. Lewis stated, however, was that it was exactly the same type of damage, but to the DNA of any body cell, that could produce cancer. This was extremely important because it had already been decisively demonstrated that genetic damage was cumulative. In one experiment after another, using fruit flies and large colonies of mice, it was found that it did not matter how slowly or quickly a given dose of radiation was administered—in every case the number of defective offspring was essentially the same. The resulting effect on offspring was determined only by the total accumulated radiation dose received, regardless of the length of the time period over which it was given. There were some indications of repair in the ova of female mice, but the effect was relatively small at best. Thus there existed clear evidence that radiation effects of the type that produced genetic damage were cumulative, especially in the male sperm cell. But if Dr. Lewis was right, and radiation caused cancer in body cells in exactly the same way as it caused genetic damage in reproductive cells, then this clearly implied that the cancer-causing effect of radiation was also cumulative.

And if this was so, then the greater the radiation dose, the greater the risk of cancer. Dr. Lewis's theory therefore supported the findings of Stewart and MacMahon, and simultaneously gave weight to the theory that the cancer-causing effects of protracted radiation from fallout would be the same as for X-rays given all at once.

All of this evidence combined pointed toward a single tragic conclusion: Man, especially during the stage of early embryonic life, was hundreds or thousands of times more sensitive to radiation than anyone had ever suspected.

3

A Small Error in the Assumptions

MY OWN INVOLVEMENT in the subject of fallout hazards began in 1961. That was the year of the Berlin crisis, when the Soviet Union ended the voluntary moratorium on nuclear testing and the U.S. government called for a large-scale fallout shelter construction program. The intensified threat of nuclear war caused much concern in the scientific community, and in Pittsburgh, Pennsylvania, a group called the Federation of American Scientists, of which I was a member, decided to participate in a study of the chances for survival of a large industrial city like Pittsburgh in the event of a nuclear war. Since I was professionally involved in research on new techniques for reducing the radiation dose from medical X-rays, and therefore was interested in the problem of low-level radiation effects, it was suggested that I join the section investigating the health hazards from fallout.

Almost as soon as we had begun our work, a disturbing fact emerged. All the calculations made by government agencies as to the radiation protection necessary after a full-scale nuclear war were based on the assumption that the adult could tolerate the enormous dose of 200 rads spread over a few days and as much as 1000 rads over a year. Apparently it had been decided by the government's scientific advisory groups that it was not necessary to take into account the long-range after-effects of radiation, either on the survivors themselves or on their offspring. Yet, as I well knew from my own research, the reason why so much effort was being spent to reduce the dose from medical X-rays was that the doses of only a few rads per year received by radiologists in the course of their work had been found to decrease their life spans significantly, while among their children there had been a definite increase in congenital defects. Furthermore, if Dr. Stewart was correct, only 1 to 2 rads would double the chances of a child developing cancer when the radiation was received in the last few months of the mother's pregnancy, and only one-tenth of this amount might have the same effect when received in the first few months. Exposed to the radiation levels that would be present in the aftermath of a nuclear war, then, a great many children born in the years following could be expected to die of leukemia, cancer, or congenital malformations before reaching maturity.

Additionally, these doses of hundreds of rads that the government agencies considered tolerable were only estimates of the external doses from the fallout in the environment. The internal doses from the fallout particles concentrated inside the body, which would be hundreds or thousands of times higher still, had not even been taken into account, although the knowledge necessary for calculating these internal doses was widely available.

It thus appeared that the chances for survival after a nuclear war were being presented to the public in a far more optimistic light than scientific evidence justified. And then, with the publication in 1962 of Lapp's article in *Science,* revealing the extremely high internal doses received by the children of Albany-Troy from the 1953 rainout, it became evident that the same held true for the health effects of peacetime fallout. Since by 1962 the intensive nuclear testing was filling the rains all over the world with radioactivity approaching the amounts that had descended on Troy, the number of children that could be expected to die as a result was very large.

I made an estimate as follows: According to figures presented at congressional hearings, the fallout from each 100 megatons* of hydrogen bombs tested would give an overall dose of from 200 to 400 millirads to every man, woman, and child in Europe, North America, and Asia. This was approximately the total megatonnage of the bombs already exploded in the latest test series as of the end of 1962, and 200 to 400 millirads was roughly equivalent to the dose from a pelvic X-ray. Thus, if there were indeed no difference in the effects of diagnostic X-rays and fallout, one could expect as much as a 20 percent increase in cancer rates for those children born within a year after the recent tests. Since, at the time, about one child in a thousand normally died of cancer before reaching adolescence, and since four million children were born in the United States each year, then every year some 4000 children normally developed cancer. Therefore, a 20 percent increase would mean close to 800 additional deaths in the United States alone. For the rest of the world, the figure would be perhaps ten times larger, all as a result of only the most recent atmospheric tests.

And these figures did not even take into account the probability of much larger doses from local rainouts, where the fallout was brought down in concentrated form. In the case of Troy, the type of calculations made by Lapp indicated that an overall dose of anywhere from a few hundred to a few thousand millirads must have been received by the unborn children in the area, equivalent to a whole series of pelvic X-rays. Depending on whether they were in an early or late stage of development at the time, their chances of developing cancer would have been increased 100 percent or more.

Thus it was clearly of the greatest importance to see whether the number of leukemia deaths among the children of Troy had in fact begun to increase a few years after the fallout arrived. (A characteristic delay in the onset of the disease, when radiation was the cause, had been found by both Dr. Stewart and Dr. MacMahon and was also observed among the survivors of Hiroshima and Nagasaki, who began developing this fatal

---

* One megaton is the equivalent in explosive energy to a million tons of TNT.

form of cancer some three to five years after their exposure.) Furthermore, it seemed imperative that the worldwide scientific community be made aware of the implications of the data of Stewart and MacMahon, and of the urgent need for large-scale statistical studies of populations exposed to fallout. Accordingly, by late fall 1962, I had completed an article on the subject and submitted it to *Science* magazine. This seemed the most appropriate place for publication, since it was the official journal of the American Association for the Advancement of Science (AAAS), the country's largest scientific professional association, and as such was read by a large, interdisciplinary audience of scientists throughout the world.

This was, however, an inauspicious time for the publication of an article with such negative implications for nuclear warfare and peacetime testing. The Cuban missile crisis, which brought the world to the brink of nuclear war, had just passed, greatly increasing pressure for further development and testing of nuclear weapons. Therefore, in anticipation of possible publication difficulties, I decided to submit copies of the manuscript to a few noted scientists in the hope of gaining added support.

One copy went to Dr. Russell Morgan, chairman of the Department of Radiology at Johns Hopkins University and head of the National Advisory Committee on Radiation of the U.S. Public Health Service. Dr. Morgan was one of the country's most knowledgeable experts in the areas of X-ray technology and low-dose radiation effects. In his reply, he stated that the article brought into focus important implications of the work of Stewart and MacMahon that had not been fully recognized, and recommended that it should be published with only a few minor changes. Dr. Morgan also gave me his permission to refer to his statement if the paper had to be resubmitted to *Science* after an initial rejection.

Another copy went to Dr. Barry Commoner, professor of botany at the University of St. Louis and one of the founders of the Committee on Nuclear Information, a group that pioneered in the public dissemination of information on the effects of nuclear testing. Dr. Commoner said in his reply: "I believe that it [the article] represents a very important contribution to the subject. I hope that it will be published in *Science* just as it stands. . . . Your conclusion regarding the need for large-scale surveys of the incidence of leukemia and other forms of cancer is of great urgency."

The article, however, was returned by *Science*, accompanied by copies of two reviews and a letter of rejection from the editor, Philip Abelson. Abelson was a physical chemist who had an extensive background in the nuclear field. For many years he had worked closely with Glenn Seaborg, later chairman of the AEC and president of the AAAS, on the development of processes for the production of uranium, and he was now a member of both the General Advisory Committee of the AEC and the Project Plowshare advisory committee. (Project Plowshare was the name given to the AEC's program for the development of peaceful uses for nuclear explosives.) In his letter, Abelson stated that he had reviewed the article himself and found that "there is not enough solid material to justify publication." He further expressed the opinion that "there is really no evidence of the functional relationship between the number of X-rays taken and cancer mortality." This meant that he did not consider significant the indications in the work of Stewart and MacMahon that the risk of cancer increased directly with the increase in X-ray dose, indications which were in sharp contradiction to the threshold theory.

Upon examining the enclosed comments of the other two reviewers, who were nameless, as is the custom, I found that one was completely negative, stating that the article presented "no new observation" and ignored studies that showed no effects from diagnostic X-rays. The other reviewer, however, recommended publication. Apparently, then, it had been Abelson's opinion that weighted the scales in favor of rejection.

A few days after the article was returned, I received an unexpected letter from Dr. James H. Lade, special assistant to the commissioner for radiological health of the New York State Health Department. Since 1951 Dr. Lade had also been director of the department's Bureau of Medical Defense, a part of the state's extensive Civil Defense Program, which had carried out an "exercise" at the time of the Albany-Troy incident.

As medical director, Lade had been one of those who participated in the decision that no health protection measures were
necessary after the incident and that no ill effects were to be
expected. His letter read as follows:

> Dear Mr. Sternglass:
>
> I have had an opportunity to review your interesting
> paper on "Ionizing Radiation in the Pre-Natal Stage and
> the Development of Childhood Cancer," and noted your
> reference to Ralph Lapp's paper on the Troy-Albany fallout
> in 1953. We in this department have done a little investiga-
> tion of the circumstances which obtained in the Troy-Al-
> bany area at that time and the number of cancer cases
> and deaths reported in the age group who were under two
> years of age at that time. You may be interested in the
> results of these investigations, summarized in my attached
> letter to *Science.*
>
> > Yours very truly,
> > James H. Lade, M.D.
> > Director

Lade, apparently, either had been the negative reviewer
or had been consulted by him. The *Science* letter to which
he referred had been published in the November 9 issue in
reply to Lapp's article. In that letter Lade attempted to minimize the possibility of any radiation effects in the Albany-Troy
area from the concentration of radioactivity in the milk by
arguing that the cattle in the area had not been turned out
to pasture until about May 12, 1953, or some two to three
weeks after the fallout had arrived on April 25. He argued
that in view of the seven-day half-life* of iodine 131, the radiation would have decreased to only one-fourth its initial intensity
by the time the cattle were turned out, so that Lapp's estimates
of the dose to infant thyroids were at least four times too high.
Lapp had, apparently, been unaware of this factor when he

---

* The half-life of a radioactive isotope is the time it takes for the radioactivity
to diminish to half its original intensity. The half-life of iodine 131 is 7 days,
so it is termed a short-lived isotope. Strontium 90, with a half-life of twenty-
eight years, is a long-lived isotope. The radiation from a short-lived isotope
is much stronger because all of it is given off in a shorter time.

made his dose estimates, but Lade failed to mention that even if Lapp's doses were reduced by a factor of four, they would still be vastly greater than the permissible limits set by government agencies. Lade's letter also did not take into account the dose from the longer-lived isotopes such as strontium 90, strontium 89, and barium 140, with half-lives of 28 years, 50 days, and 13 days, respectively, which would certainly still be present two to three weeks after the fallout had arrived. Nor did he mention the information published by Professor Clark in the *Journal of the American Water Works Association,* namely, that in May and June, many weeks after the first rainout, fallout from additional tests repeatedly produced new levels of radioactivity comparable to those measured for April. This meant, of course, that Lapp's estimates, far from being four times too high, were actually much too low, for Lapp had based his estimates only on the April 26 fallout. Lade, as medical director of the Civil Defense group, worked intimately with the scientists from the New York office of the AEC who sponsored the measurements of radioactivity in the reservoirs of the area. Thus he presumably would have been aware of this circumstance.

Lade further argued that because of the heaviness of the spring rains, the radioactivity had soon been largely washed off the vegetables and pasture. But Dr. Clark and his students had found that even extreme chemical treatment was only partially effective in removing the radioactivity from the leaves and other objects to which it clung. Thus, ironically, the only effect of the rains of May and June would have been to bring down even greater amounts of fallout than had come down in April.

As a final point, Lade stated that a review of New York Health Department records indicated that no cases of thyroid cancer had since developed among the children who were under two years of age in 1953. His letter closed with the remark that "it seems most unlikely that an event which has resulted in no increase of thyroid carcinoma during the ensuing nine years will lead to such an effect in the future." Yet it was common knowledge among specialists in the field that radiation-caused thyroid cancer generally takes ten to twenty years to develop. And Lade said nothing about any increase in the incidence of leukemia, which by this time would certainly be detectable.

It could thus be determined by someone with a scientific background that Lade's letter contained absolutely no evidence to support his conclusion that the Troy fallout had been harmless. But how could the general public ever guess? This was the voice of the New York State Health Department.

Within a few weeks, I resubmitted my article to *Science* together with a letter referring to the statement given me by Dr. Russell Morgan. Within less than a month, it had been accepted for publication.

4

A Ray of Hope

BY THE SPRING of 1963, when the article finally appeared in *Science,* the levels of radioactivity in milk were reaching unprecedented heights all over the United States. Extreme concern was being voiced both by scientists and the general public regardless of the repeated reassurances by local public health officials and the AEC that no danger existed. The intensified pressure began to penetrate to the highest levels of government, and in June 1963 President John F. Kennedy announced that this country, the Soviet Union, and Great Britain had agreed to negotiate a treaty to end all atmospheric testing. He further stated that he had ordered an end to all such testing by the United States. In July the President delivered an address to the nation in which he urged the ratification of the treaty by the U.S. Senate. In this address, he referred to the threat of fallout as follows:

> . . . the number of children and grandchildren with cancer in their bones, with leukemia in their blood, or with poison in their lungs might seem statistically small to some, in comparison with natural health hazards, but this is not a natural health hazard—and it is not a statistical issue.

The loss of even one human life, or the malformation of even one baby—who may be born long after we are gone—should be of concern to us all. Our children and grandchildren are not merely statistics toward which we can be indifferent.

It appeared that the issues involved in the question of fallout hazards were at last receiving widespread public recognition. In Congress, hearings on low-level radiation effects were being held by the Joint Committee on Atomic Energy, and for the first time citizens' groups and private scientists with no government affiliations were invited to testify on this subject. In August I received a letter from John T. Conway, executive director of the Joint Committee, requesting my presence at the hearings. Leafing through the transcript of the first half of the hearings held in June, I found a reprint of my *Science* article. It was followed by a lengthy critique prepared by the AEC's Division of Biology and Medicine. On examination, the AEC critique proved to be strikingly similar in wording and theme to the negative review for *Science.* It denied that any effects from low-level radiation had been proven to exist, stating that I had ignored the studies which showed no effects. The AEC critique, however, went further and cited figures showing an actual reduction in childhood cancer rates after the heavy atmospheric testing in 1957. But the figures were only for children who were less than two years old when they died, a peculiar restriction, since Stewart and MacMahon had specifically demonstrated that cancer caused by the irradiation of unborn children only showed up *after* the age of two. No figures for older children were given by the AEC.

The Joint Committee hearings opened on August 20, the same day that Dr. Edward Teller testified before the Senate against ratification of the Test-Ban Treaty. Dr. Teller, the renowned nuclear physicist who had been instrumental in the development of the hydrogen bomb, expressed in his testimony the main arguments advanced by the treaty's opponents. He raised the possibility of future military breakthroughs by the Russians, stating his conviction that in the 1961–62 test series they had acquired knowledge about missile defense that this country didn't have. "This is the main argument against the test-ban treaty," he said. "It weakens our defense, and as long as we have reason to distrust Soviet intentions, the weakness of our defense will invite attack. . . . I came to recognize many years ago that real cooperation between Russia and us in the near future is impossible." As to the possibility of danger from fallout, Teller flatly stated: "From the present levels of world-wide fallout there is no danger. The real danger is that you will frighten mothers from giving milk to their babies. By that, probably much more damage has been done than by anything else concerning this matter." Teller also stated that the treaty would endanger the program for the peaceful uses of nuclear explosions in the Project Plowshare program. He expressed his belief that such explosions could be carried out in "a very clean way," so that fifteen minutes afterward people could walk in the resulting crater "without exposing ourselves to more radiation that we have taken year in and year out in our laboratories."

At the same time as Dr. Teller was making this statement, in another part of the capital Dr. C. W. Mays of the University of Utah was testifying at the Joint Committee hearings about the thyroid doses received by children in Utah from one of the AEC's Project Plowshare explosions the year before. As measured by University of Utah scientists, the doses ranged from 10 to 60 rads—anywhere from ten to a hundred times the government's maximum permissible yearly limit for that body organ.

The second day of the Joint Committee hearings, during which I was present, began with the testimony of Dr. Shields Warren, who represented the National Academy of Science's Advisory Committee on the Biological Effects of Radiation. Dr. Warren had been the first director of the AEC's Division of Biology and Medicine, and as such he had been responsible for all of the AEC's early research and planning relating to the health effects of fallout. He was also head of the U.S. delegation to the United Nations Scientific Committee on the Effects of Radiation. Dr. Warren began his testimony with a review of the effects of radiation on man, animals, and plants. The lowest dose listed on his chart was 0.001 rad, or 1 millirad, and this was followed by 0.01 rad, which he indicated as giving "no detectable effect." The next highest dose was 1.0 rad, again with the legend "no detectable effect." Dr. Warren evidently did not accept the evidence of Stewart and MacMahon, which indicated that 1.0 rad received during pregnancy produced something like a 50 percent increase in the rate of childhood leukemia and cancer, clearly a "detectable" effect.

Nothing very serious appeared on Dr. Warren's chart until he got to 1000 rads. At 10 rads the legend was: "Barely detectable qualitative changes in lymphocytes," while at 100 rads the chart indicated only "mild acute radiation sickness; slight diminution in blood cell counts. Possible nausea and vomiting. . . ." Yet the studies of Hiroshima had revealed a doubling or tripling of leukemia deaths among the surviving adults exposed to this dose.

Only at 1000 rads did radiation actually appear lethal: "Depression of blood cells and platelets . . . death within twenty days." This was profoundly misleading: It was common scientific knowledge that 50 percent of all individuals exposed to 400 rads to the whole body would die within a matter of weeks or months, while for 700 rads the figure was 95 percent.

For 10,000 rads, Dr. Warren's chart listed "Immediate disorientation and coma. Death within hours . . ." while "death of all living organisms" was reassuringly indicated as not occurring until 10,000,000 rads. In conclusion, Dr. Warren quoted the following passage from the United Nations 1962 report on fallout hazards, which he had helped to prepare:

> It must be recognized that the human species has, in fact, always been exposed to small amounts of radiation from a variety of natural sources and that the present additional average exposure of mankind from all artificial sources is still smaller than that from natural sources.

This, then, had been the voice of the AEC's Division of Biology and Medicine, the National Academy of Sciences, and the U.S. delegation to the United Nations Scientific Commission.

When Congressman Melvin Price asked Dr. Warren what effects he would expect to observe in the 250,000 children in

Utah who might have received 4.4 rads to their thyroids, as Dr. Mays had testified the day before, he replied: "I think it would be hard to find any effect, Mr. Chairman." But as Dr. Mays had suggested, if one does not look, one is not likely to find such effects, and indeed, no one had conducted the necessary studies of the children in Utah, in Albany-Troy, or, for that matter, anywhere else in the world.

The next witness was Dr. E. B. Lewis of the California Institute of Technology, whose "single-hit," linear theory of cancer causation supplied an explanation for the findings of Stewart and MacMahon. This theory had formed the basis of my argument that the cancer-causing effects of radiation were cumulative and that doses received gradually from fallout would be similar in effect to those received all at once from diagnostic X-rays. Dr. Lewis's latest evidence, based on a study of leukemia incidence among radiologists, indicated that low doses of radiation delivered over a period of many years could indeed lead to an increased incidence of leukemia, a finding that further buttressed his original theory.

There was now only one more witness scheduled to be heard before Dr. Brian MacMahon. This was Dr. Hyman Friedell, professor of radiology at Western Reserve University. The main burden of his testimony, as he stated, was to urge radiation protection agencies to set their standards on the basis of the linear theory of radiation damage, according to which there was no safe threshold of exposure.

Now Dr. MacMahon took the stand. He opened his testimony with an explanation of why certain other studies had not appeared to confirm Dr. Stewart's and his own results: "Every one of the negative studies has been based on small numbers, and in no instance do the results differ in a statistically significant degree from the expectation of a 40 percent increase in cancer risk among exposed children." Thus, in defense of his own work, Dr. MacMahon had effectively answered one of the major points in the AEC's critique of my article. He went on to summarize his results as follows:

I suggest therefore that the existing evidence is overwhelmingly indicative of an increase in cancer risk for children diagnostically exposed in utero; that this increase is seen for leukemia as well as for a variety of other individual forms of cancer; that the best estimate of this increase is that it is about 40 percent in excess of "normal" cancer mortality in the United States.

Turning next to the implications of his findings, he stated:

It seems to me that if this association is accepted, we must consider very seriously the possibility of cancer production by low doses of radiation such as encountered in X-ray diagnosis and even fallout.

As to the existence of a possible "threshold," Dr. Mac-Mahon went on to say:

The exposure dose associated with pelvimetry [pelvic X-rays] at the time of these studies is not known, but it was probably of the order of 2 to 3 rad [for a series of X-rays]—substantially below the 50 rad that has frequently been mentioned as a possible threshold level. If a threshold for leukemia induction exists, then it must be below 2 rads. It would be a coincidence indeed if a threshold existed just below the dosage level at which studies have been undertaken.

One other major issue remained to be discussed, and that was the evidence for an increase in cancer risk with an increase in dose—termed the "dose-response" relationship. To this point, MacMahon now addressed himself as follows:

Furthermore, there is some indication in both Stewart's and our own data that a dose-response relationship exists even within the low-dose range that is being considered.

Thus, MacMahon, together with all the other nongovernment scientists who testified in this part of the hearings, presented evidence in conflict with the AEC's claims for the existence of a safe threshold.

Inevitably, Dr. MacMahon was asked his opinion of the use to which I put his data in my *Science* paper. He summarized his reply as follows: ". . . the argument used by Dr. Sternglass does not convince me personally, but, on the other hand, I cannot deny that ultimately his point of view may be correct." He further stated, however, that he agreed with all the comments in the AEC critique. This seemed puzzling, since I felt he had just answered the most serious points in his own testimony. It became clear that I would have to amass much stronger evidence before my hypothesis could gain the public support of many scientists.

After a brief recess, Dr. Eric Reiss was called to the witness stand. Dr. Reiss was a St. Louis physician and a co-founder, with Dr. Barry Commoner, of the St. Louis Committee for Nuclear Information. In his presentation on behalf of the St. Louis group, he concentrated on the subject of local fallout incidents resulting from the AEC's tests in Nevada. Referring to the many reports on local fallout prepared by the AEC, he pointed out that the general conclusion in all these official reports had been that "the test program has been carried out without any discernible threat to the safety of the local population."

"In contrast," Dr. Reiss went on to say, "our analysis of the same monitoring data published by the AEC shows that as a result of nuclear tests at the Nevada Test Site, in the period 1951–62, a number of local populations, especially in Nevada, Utah, and Idaho and probably other communities scattered throughout the continental United States, have been exposed to fallout so intense as to represent a medically unacceptable hazard to children who may drink fresh locally produced milk." He explained that the discrepancy between the AEC's conclusions and those of the St. Louis group arose out of the fact that the AEC only measured the external dose from the fallout in the environment, while his group also measured the internal dose from the fallout particles concentrated inside the body, as Dr. Lapp had done in the case of Troy.

Dr. Reiss proceeded to present figures indicating that for 10 rads to the thyroid approximately one in 2860 children might be expected to develop thyroid cancer. Because of the long latent period for this disease, he explained, the damage would not show up for many years, possibly ten to twenty.

He then cited estimated thyroid doses of 3 to 18 rads for the large population of Salt Lake City, Utah, as a result of fallout on May 7, 1952, and 5 to 40 rads for Albany-Troy on April 26, 1953 (this did not even include the much heavier fallout in June). Such hazardous doses, Dr. Reiss pointed out, arose not only from atmospheric tests but also from certain underground tests: "Venting [leaking of radioactivity] has been reported for at least seven underground nuclear tests," said Dr. Reiss, adding that his group was able to calculate, for example, that "fallout from the underground 'Gnome' shot delivered sufficient fallout to the vicinity of Carlsbad, New Mexico, to cause thyroid dose levels of from 7 to 55 rads to children."

And then Dr. Reiss turned to an astonishing aspect of the whole problem: If the dangerous internal doses had been estimated at the time, "simple preventive measures could have been taken to avoid exposure." It would only have been necessary to warn the inhabitants of the area to avoid drinking fresh milk produced locally until the iodine levels died down. "We know of no instance," said Dr. Reiss, "in which such a warning was issued until the summer of 1962, when high iodine 131 levels observed in commercial milk supplied in Utah led state health officials to divert current milk from the market." And finally, Dr. Reiss noted that correct estimates of internal doses should have been possible by as early as 1954 *on the basis of the then-available scientific theory.*

The questioning of Dr. Reiss was nearing its end when John Conway asked him why his group had made only theoretical calculations of the thyroid doses without benefit of any direct measurements of the iodine levels in the milk. "I am delighted you brought that up," Reiss answered, "because it is the nub of the problem. If measurements were available, we would obviously have used them. The question is, why were no measurements made or reported publicly by those who had been charged with the protection of our health and safety?"

That evening the *Washington Post* gave prominent coverage to Reiss's testimony, as well as to the testimony given at the Senate's test-ban ratification hearings. At those hearings, Dr.

John S. Foster, Jr., the new director of the AEC's Livermore Laboratory, where the hydrogen bomb had been developed, had offered his opinion on fallout hazards:

> You are asking for a quantitative answer which involves the understanding of an effect, namely the effect on the human being that is so small compared with his [radiation exposure from] background that we have not been able to measure it. One way to look at it would be to say that the fallout from all past tests affecting man for the next 50 to 100 years would be something like the same thing as deciding to live a few hundred feet higher up, higher above sea level; that is what it means.

Senator Robert Byrd of West Virginia then asked Dr. Foster what importance should be attached to the public's fears of fallout. Byrd phrased his question as follows: "If I may pose a hypothetical question, are you saying, in essence, that if you were a senator with the knowledge that you possess in the scientific field . . . you would disregard entirely this factor [public opinion] in your reaching a decision?"

To which Foster replied: "That is correct, sir, although it would be a very difficult position for me to put myself into. . . ."

The third day of the Joint Committee's low-level radiation hearings opened with my testimony. I reviewed all the evidence regarding low-level radiation effects on unborn children, presented my arguments as to the probable similarity in effects of X-rays and fallout radiation, and gave estimates of the increase in childhood cancer that might be expected from nuclear testing. Then the questioning period began.

Congressman Craig Hosmer asked me how there could be any evidence in Stewart's and MacMahon's studies for a dose-response relationship when the radiation dose from hospital X-ray machines was known to differ widely from hospital to hospital, due to variations in the adjustment and quality of the machines. Such differences would mean that some mothers might have received the same or greater dose from a single X-ray as others received from two or even three. Therefore, the finding of Stewart and MacMahon that there was an increase in cancer risk with the number of X-rays taken might not indicate a dose-response relationship. I explained that it was also true, however, that in the case of any one machine, the mothers who received two X-rays from it would definitely have received twice as much radiation as those who were X-rayed only once. Since the ratio of single pictures to double or treble pictures taken was generally the same for each X-ray machine, there could be no question that, on the average, the mothers who received two X-rays had received twice as much radiation as those who received only one.

Congressman Hosmer next asked why one couldn't test the validity of the dose-response relationship simply by examining the statistics for leukemia in children after the test series in the early 1950s. I pointed out that this was precisely what Lapp, Mays, Reiss, myself, and others had urged the government to do without success.

The next question involved the principal argument used by the AEC to minimize the possible effects of fallout. Congressman Holifield asked whether I was aware that testimony had been presented by government scientists that fallout from all past testing had raised background radiation by only some 10 percent. Since the typical background dose was 100 millirads per year, this seemed to be only 10 millirads additional radiation, much less than the 200 millirads received from a modern pelvic X-ray, or than the dose of 200 to 400 millirads that I had calculated for the recent test series.

What the spokesmen for the AEC and the Federal Radiation Council had done was to take the total radiation dose from all the bombs detonated so far—some 700 millirads—and average it out as if it were received uniformly over a seventy-year life span at a rate of 10 millirads per year. But in fact, most of the dose comes from the short-lived isotopes that predominate in fallout, and is thus delivered in the first few months after the tests. To illustrate this, I cited figures published just a few months before by the AEC's Brookhaven National Laboratory, located on Long Island, not far from New York City. According to these figures, during the first six months of 1963, the average radiation exposure on Long Island had risen to 4.7 millirads per week, as compared with a normal background rate of only 1.7 millirads per week. This meant that in only half a year the dose received by individuals in this area would be about 122 millirads as compared with a normal dose of only 44 millirads. Thus, instead of the negligibly small 10 percent increase which the AEC's seventy-year averaging procedure would give, the actual exposure to an infant in the womb would have been nearly triple the background rate. And this dose did not even include the effects of internal concentration. This, then, was the reasoning behind the reassuring statements of the world's radiation protection agencies that doses from atmospheric testing were negligible compared with all the sources of radiation in man's natural environment.

Congressman Hosmer next raised the point that the greater background radiation due to cosmic rays in an area of high altitude such as Denver as opposed to that in an area at sea level such as Long Island should lead to a higher leukemia rate in Denver. But in point of fact, as I stated, the difference in total background radiation between these two areas is actually quite small because most background radiation comes from sources other than cosmic rays. Cosmic rays contribute only about 30 millirads at sea level, compared to 50 millirads from the rocks and soil and 20 millirads from traces of natural radioactive substances in our body. Going to an altitude of 5000 feet increases the cosmic ray dose by only 40 percent, or a total of 12 millirads. Since place-to-place differences of 20 or 30 millirads in the radioactivity of rocks are common, the small difference due to altitude is easily masked by this factor alone. Furthermore, cosmic rays do not concentrate in critical body organs as does fallout.

Thus, it would be nearly impossible to find differences in leukemia rates as a result of altitude differences even as large as those for Denver and Long Island, considering that Mac-Mahon had to use a study group of some 70,000 exposed children to clearly establish an effect from about 1000 millirads of X-rays. And I added that there are of course other factors beside radiation that enter into the likelihood of a child developing leukemia, making such a comparative study of the effects of slightly different background levels next to impossible.

Following up this point, Congressman Hosmer suggested that if factors other than radiation were involved in the cause of leukemia, did this not invalidate MacMahon's study? To this I replied that, in such a study, comparisons are carried out with control groups for whom all the other factors are, on the average, nearly the same as for those exposed to X-rays, thus essentially eliminating the effect of such factors on the outcome.

After some discussion of the size of the dose from fallout, the chairman indicated that the committee's counsel, John Conway, had some points he wished to raise. After a lengthy prelude, Conway led up to the fact that the Federation of American Scientists' news release based on my paper said that 100 megatons of nuclear fission energy had been released in the 1962 test series. Conway argued that since the amount released in 1962 was actually 76 million tons, then my estimate of the expected number of leukemia deaths was too high.

But it was clear from both my original paper in *Science* and my written testimony that the figure of 100 megatons applied to the entire test series that began in September of 1961 and ended late in 1962, not just to the amount detonated in 1962 alone. Thus, the news release, which was not even prepared by me, should have said more precisely, "1961–62 test series."

When this became clear, Chairman Price intervened to point out the simple nature of the misunderstanding, and when Conway nevertheless wanted to pursue the matter further Price thanked me for my testimony and called the next witness.

The debate over the test-ban treaty continued in Congress through most of September, and then the U.S. Senate voted overwhelmingly for ratification.

A highly revealing postscript to the entire test-ban treaty debate came to light in December, when the full proceedings of the Joint Committee hearings were published. Inserted in the record was a letter from Dr. Harold Knapp, a fallout specialist who had recently resigned from the AEC's Division of Biology and Medicine. In his letter, Dr. Knapp made reference to a 1957 incident in which a nuclear reactor in Windscale, England, had gone out of control and emitted tremendous quantities of radioactivity into the atmosphere. The contamination from iodine 131 was so great that the crops and milk for hundreds of square miles around had to be seized and dumped. According to Dr. Knapp's letter, in 1960, when he was doing fallout research for the AEC, he came to the startling conclusion that in Utah "depositions of iodine 131 per unit area *at many inhabited ranches and communities must on several occasions have exceeded the maximum iodine 131 concentrations on pasturage found after the Windscale accident.*" (Emphasis added.)

As an example, Knapp cited an incident of relatively heavy fallout in the St. George, Utah, area on May 19, 1953. His calculations indicated that 24 hours after the explosion the iodine levels in milk must have reached 700,000 to 2,600,000 micromicrocuries per liter. He estimated that the thyroid dose for an infant who drank one liter (slightly more than a quart) of this milk each day for the three weeks following the test would be anywhere from 120 to 440 rads.

Knapp wrote a report containing these conclusions in 1960, but the AEC did not release his report for publication until August 16, 1963, just a few days before the second half of the Joint Committee hearings began, by which time it was evident that the independent scientists from Utah and St. Louis were going to make public their own similar findings. Thus, during the entire early effort to achieve a test-ban treaty, this shocking and vital information was kept from the people and political leaders of the world, while hundreds of megatons of bombs were exploded by the U.S. and Russia during 1961–62. And since Knapp's report had not even been made public by the AEC and the Joint Committee by the time of the first half of the low-level radiation hearings in June of 1963, it had still been possible for the AEC to mislead Congress and the public until just before the end. From page 225 of the proceedings of the June hearings:

REPRESENTATIVE HOLIFIELD: And the testimony before this committee has been that tolerable permission dose has not been reached by the amount of radiation that

comes from manmade fallout radiation, is that not true?
DR. HARLEY: Yes, Sir.

Dr. Harley was John Harley, the former colleague whom
Professor Clark had telephoned at the time of the Albany-
Troy incident. At that time a scientist with the AEC's Health
and Safety Laboratory in New York City, Harley had since
become head of the lab, which had been responsible for the
classified reports on the Troy fallout during the spring of 1953.

5

The Evidence Begins to Emerge

IT WAS NOW an established fact that potentially dangerous radiation doses had been received by large numbers of people from fallout. What remained to be established was whether or not these doses had actually caused any damage. No direct statistical evidence had yet been gathered by either the AEC or federal and state health agencies, or, if it had been gathered, it had not been published.

This situation was abruptly transformed in March of 1964, when another letter by J. H. Lade of the New York State Health Department appeared in *Science*. This letter was written in response to repeated prodding by Ralph Lapp, who kept challenging the department to make public the actual figures after Lade had claimed in the September 1963 issue of *Science* that no increase in childhood leukemia had occurred as a result of the 1953 rainout. The exact words used by Lade in making this claim had been:

The cancer report files of this department reveal no increase in the incidence of cancer or leukemia over the past ten years in children of the Albany, Troy, and Schenectady areas—who were 15 years or younger in 1963—as compared with children of this age elsewhere in upstate New York.

Lade's new letter contained the first detailed information on leukemia deaths among the children of the area. This was the data on which he had based the above conclusion.

The data consisted of a table that included all reported fatal leukemia cases among children under fifteen years of age in the Albany-Troy-Schenectady area between 1952 and 1962, together with the years in which these children were born. Examination of the table showed that, beginning in the fourth to fifth years after the 1953 rainout, the yearly number of reported leukemia cases quadrupled. This was strikingly similar to the delay in onset observed in the studies of Stewart and MacMahon and among the survivors of Hiroshima and Nagasaki. In the Troy area, after the eighth year, the number of cases began to decline once more, as they also did in the two Japanese cities. During the years 1952–55, before any radiation-caused leukemia cases would be expected to appear, there were a total of nine cases among children under age ten for whom the data was complete. During the years of expected peak incidence, 1959–62, there were a total of thirty cases. Statistical estimates demonstrated that the chances were less than one in a billion that this increase of twenty-one cases could be regarded as purely accidental.

And when the cases were examined according to the year of birth, there was a very noticeable sudden increase in leukemia among the children born in 1953 and 1954, the period when the fallout radiation would have been strongest. When grouped in two-year periods, the figures in Lade's table for children under ten showed the following:

Birth Years	No. of Cases
1943–44	4
1945–46	5
1947–48	8
1949–50	5
1951–52	6
1953–54	13

How, then, had it been possible for Lade to claim that there had been "no increase in the incidence of cancer or leukemia over the past ten years in the children of the Albany, Troy, and Schenectady areas"? Upon rereading his exact statement, it became apparent that he had used a modifying phrase whose significance would easily escape the casual reader. This phrase was: "as compared with children of this age elsewhere in upstate New York."

But the "classified" measurements made by the AEC's New York lab had shown that the *entire upstate New York area* received heavy fallout on April 26, 1953, and again during the next few months. Under these circumstances, there would of course be little or no *difference* in leukemia rates between Albany-Troy-Schenectady and elsewhere in upstate New York. Although he presented no data, Lade's statement clearly implied that in fact there had been similar increases in *all* these areas. (And indeed, later investigations revealed that there had been such rises all over New York State following three to five years after the onset of Nevada testing.)

In his 1964 letter, Lade presented a final argument to "rule out the fallout as a major factor in the leukemia incidence in that area for the 1953 births." According to his figures, there were also a large number of leukemia cases among the children born in 1956. Thus, Lade argued that the sharp rise in leukemia among those born in 1953 could not have resulted from the fallout, since there was also an increased incidence among the children born in 1956, when there presumably was no major fallout incident.

This piece of evidence seemed puzzling, especially since all the other data pointed so strongly to fallout. But then, a few months later, new light was thrown on the situation when the news of an extremely important study appeared in *The New York Times*. A group of investigators working with Dr. Saxon Graham at the Roswell Park Memorial Cancer Research Institute in Buffalo, New York, and at two other cancer research institutions, had found that the children of parents who had been X-rayed as many as five to fifteen years *before* the mothers became pregnant also had a significantly increased likelihood of developing leukemia. This finding suggested that radiation could cause a type of *genetic* damage to the sperm cells of men and the ovaries of women that would make their children more prone to developing this disease.

In the case of Troy, then, it was indeed possible that such a genetic effect of radiation could lead to a greater incidence of leukemia among the children born in the years following.

There was another very important finding of the Roswell study: Children who had received multiple exposures from a series of diagnostic X-rays to different parts of their bodies had a distinctly greater risk of leukemia than those who had only one part X-rayed. Fallout, due to the concentration of isotopes in various parts of the body, would lead to exactly this type of multiple exposure. Thus, in the Troy area, even children who were born *before* the rainout would be expected to show an increased risk of leukemia.

Independent studies had now demonstrated that radiation could cause leukemia in children prior to conception, in the womb, and after birth. What would the Troy data show if the cases were examined according to which of these three stages of development the children were in when the fallout arrived? When this was done, a fact of utmost significance emerged: Those children who had been in the womb or who were already born at the time of the fallout were, on the average, one or two years older at death than the children who had been conceived afterward. This was exactly the same characteristic shift toward older age at death that had been observed by both Stewart and MacMahon in their studies of children X-rayed while in the womb, and it had also been observed in studies of leukemia among infants who had received medical X-ray treatment. It was as if the children of Troy had been marked with a sign. The fact that the children conceived after the fallout showed the normal earlier age at death was perfectly consistent, for it is generally acknowledged that there is a large genetic factor in all normally developing leukemia cases and that a significant portion of these cases are caused by genetic damage from natural background radiation. Thus the children whose leukemia had been caused by radiation damage to their parents' genes would be expected to show the normal age at death.

None of the data supplied by Lade was therefore inconsistent with the hypothesis that the fallout had caused the increase in leukemia in the Albany-Troy-Schenectady area. And this might well be true for other types of childhood cancer. Further investigation was definitely indicated. Since I anticipated difficulty in obtaining more complete and up-to-date data from Lade, I again contacted Dr. Russell Morgan, chairman of the U.S. Public Health Service's National Advisory Committee on Radiation. After reviewing Lade's data and my conclusions, Dr. Morgan suggested that I use his name in writing to the New York State Commissioner of Health, Dr. Hollis Ingraham. In July I wrote Dr. Ingraham, asking for further information relating to cancer incidence in the Troy area and mentioning Dr. Morgan's concern about the data published by Lade. A few weeks later I received a letter from Lade indicating that he felt sufficient data was already available in his 1964 *Science* letter. It appeared that the New York State Health Department did not want to see the matter pursued any further.

6

The Hidden Tragedy
of Hiroshima

LADE'S DATA on leukemia among the children of the Troy area, published in the expectation that it would disprove the existence of any effects from the 1953 fallout, now seemed to prove the very opposite. But there remained one major difficulty in accepting the evidence for an increase in leukemia, and that was the fact that the studies of the children conceived after the atomic explosions in Hiroshima and Nagasaki in 1945 had shown no such increase in leukemia, or in any other radiation effects, for that matter.

There had, of course, been very serious effects on the infants who were in the womb at the time of the explosions. A large fraction of these infants were lost through premature death, while among the few that survived more than a year, many suffered from congenital malformations and mental retardation. Of those unborn children who had been about one mile from the explosions and received estimated doses of 10 to 20 rads, nearly a third were found to have reduced brain size and mental retardation.

But among the many children *conceived* after the explosions, there appeared to be no effects, as opposed to the situation in Troy.

Late in 1966, I was given a copy of the preliminary results of a Yale University study of Hiroshima and Nagasaki carried out by Dr. S. Finch and a group of colleagues and sponsored by the Atomic Bomb Casualty Commission. The study involved an examination of the incidence of leukemia among some 17,000 children whose parents had been within 2000 meters of the explosions. Radiation doses ranged from about 3 to 30 rads, with a few as large as a hundred. These were certainly much larger than the doses from typical pelvic X-rays received by the mothers in the study by Dr. Saxon Graham, doses which increased the risk of childhood leukemia among the children conceived many years later by as much as 100 percent.

The control group used for comparison purposes in the Yale study consisted of the population in the suburbs farther than 2500 or 3500 meters from the explosion, where the radiation from the bomb was calculated to have been less than that from natural background. With such a tremendous difference in doses, there could be no question that there should be a substantial difference in effects. Furthermore, since there had been virtually no fallout in the two cities, the doses had been calculated from the instantaneous flash of the bombs alone. Such a flash was similar to the way in which the dose from diagnostic X-rays is delivered, so, if anything, the results of Dr. Graham's study should have been even more applicable to this situation than to Troy, where the fallout dose was delivered over a period of months or years.

Still another study of Hiroshima and Nagasaki had also found a similar lack of effect. This study had been conducted over a period of nearly two decades by Drs. J. V. Neel and W. J. Schull of the Department of Human Genetics at the University of Michigan. The puzzling nature of the results was emphasized by Dr. Neel himself during the course of a lecture series in 1963: "In view of the vast body of data regarding the mutagenic effects of radiation, it can scarcely be doubted that the survivors of Hiroshima and Nagasaki sustained genetic damage." To this he added: "The question is not 'Is there damage?' but rather 'Can the damage be detected?' "

It was clear that, unless this problem could be resolved,

any evidence on the Troy incident would always be subject to this serious criticism. Accordingly, one day I decided to look up the original data on the radiation dose measurements for the two Japanese cities, collected by E. T. Arakawa of the AEC's Oak Ridge National Laboratories. And as I examined Arakawa's figures, I noticed that while there had indeed been little fallout in Hiroshima and Nagasaki proper, the fallout had drifted down on the suburbs a few miles away.

But if there had been fallout in the suburbs and beyond, and this fallout was far more damaging to the ova, the embryo, and the infant than anyone had realized when the studies were set up, then the so-called nonexposed control populations beyond about 3000 meters, the groups used for comparison purposes, had in fact *also* been affected by the radiation. And just as in the case of Lade's comparison of the leukemia rate in the Albany-Troy-Schenectady area with that in nearby areas of upstate New York, a search for a difference would fail to indicate any effect.

The fallout doses were by no means small. Arakawa reported that in the western suburbs of Hiroshima, namely Furue, Takasu, and Koi, about 5000 meters from the blast, the external radiation dose from the fallout in the environment alone probably amounted to several rads—at least twenty times what Dr. Clark and his students were to estimate for the Troy population eight years later. In the case of Nagasaki, the fallout had been even heavier, giving external doses as high as 100 rads in the suburb of Nishiyama. This was a thousand times greater than in Troy. Such doses would greatly increase the leukemia rate for the children born all over the nearby regions that had been supposedly free of radiation, completely masking the effect if one looked only for a comparative difference, and not for an increase over the rate for preceding years. It seemed incredible that the possible effects of such large doses could have been overlooked in the two major studies of the effects of atomic warfare, yet they had been.

A few months later, in the spring of 1967, after I had taken up my new position as professor of radiation physics at the University of Pittsburgh School of Medicine, I came across the evidence that confirmed this conclusion. It was contained in a paper published in the *New England Journal of Medicine,* bringing up to date all the findings on leukemia among the survivors of the two Japanese cities. One glance at the first figure told the story. As compared to the rate for the preceding years, not only had there been a rise in leukemia incidence among those who were less than 1500 meters from the explosions, but there was a similar though somewhat smaller rise for the population beyond 10,000 meters who could not possibly have received any of the direct radiation from the flash of the bomb. Both rises followed the characteristic pattern of radiation-caused leukemia. Furthermore, there was another sharp rise in leukemia for both of these groups in 1958, four to six years after the first large series of hydrogen bomb tests in the Pacific and Siberia in 1952–54, tests that resulted in heavy fallout in Japan and in contamination of the fish used widely as a staple item in the Japanese diet. Again the delay in onset coincided exactly with the established pattern of radiation-caused leukemia.

In fact, for all of Japan, leukemia rates rose sharply by 50 percent between 1946 and the early 1950s, just as Dr. Stewart's statistician, David Hewitt, had originally observed for England and the United States. This was followed by another sharp rise as of 1959. And just as the rates had turned down again in Hiroshima and Troy, they declined again throughout Japan to half their peak intensities four to six years after the temporary moratorium of 1958–61, proving that these rises had not been due to such factors as improved diagnostic methods or increased use of medical X-rays, as some had suggested.

The last major argument against the connection between the 1953 rainouts and the rise in leukemia in Albany-Troy had disappeared. By now, some three years had elapsed since my last attempt to get additional data on the Albany-Troy-Schenectady area from the New York Health Department. I sent another letter to Dr. Lade in a final effort to obtain more detailed and up-to-date information. Within a few weeks, the following reply arrived:

Doctor Sternglass, Sir:

I would be most willing to provide you with the data you request in respect to the occurrence of leukemia in children in the Albany-Troy-Schenectady area if there were any reason to suppose that they had sustained a significant exposure to fallout radiation. In my letter to the editor of *Science*, 141, 1109 (13 September 1963), I pointed out that children on a milk diet in this area at the time of 1953 could not have had a significant exposure. I fail to see, therefore, how further data could be "valuable for our understanding of low dose rate effects."

JHL

Reluctantly, I began the preparation of a final paper for *Science*, based on the incomplete data from Lade's brief letter of 1964.

It appeared unlikely that *Science* would publish the Albany-Troy paper in the near future, so I decided to present my findings at the forthcoming annual meeting of the Health Physics Society, to be held in Denver, Colorado, in June of 1968. This professional society had been founded in 1955 by Dr. Karl Z. Morgan and a few other physicists who, since the early years of atomic energy, had been concerned with the health aspects of this technology. The society was officially dedicated to "the protection of man and his environment from unwarranted radiation exposure." Over the years it had acquired many members who were professionally engaged in safety planning for nuclear weapons tests and nuclear industry activities.

Dr. Morgan, one of the most widely respected individuals in the health physics field, had himself become a controversial figure in recent years, due to his outspokenness regarding the widespread use of inadequate medical X-ray equipment and procedures. Long convinced that there existed no evidence for a completely safe "threshold" of radiation exposure, Dr. Morgan had incurred heavy criticism from radiologists' associations and others for his estimates that thousands of additional deaths of unborn children were being caused each year by unnecessary overexposure to medical X-rays. Prior to the June meeting, I sent Dr. Morgan a copy of the paper I had just submitted to *Science* and received a letter from him expressing deep concern about the seriousness of the evidence, as well as indignation about the refusal of the New York State Health Department to provide the additional data.

At the annual meeting in Denver, the press was well represented, word having gotten around about the nature of my findings. I presented the evidence and concluded by urging that much more detailed studies be made of other areas known to have received heavy fallout. During the discussion period that followed, many of the questions dealt with the inadequacies of Lade's data—for example, the lack of such facts as the birthplace of each case and the particular type of leukemia involved. I could only point out that this information was not available to me.

Reports of the Troy findings were carried widely by the press both in the U.S. and around the world. Among the many phone calls that came during the aftermath was one from an Associated Press reporter in Detroit, who informed me that the New York State Health Department had just issued a news release stating that "there is no evidence to support a Pittsburgh professor's report that radioactive fallout over the Albany-Troy, N.Y., area in 1953 has increased the incidence of childhood leukemia." The subsequent AP dispatch based on this release added that "Sternglass had consulted the department earlier and had been told the department's opinion." This "consultation," of course, had consisted only of requests for information that were repeatedly denied, while "the department's opinion" consisted of the arguments advanced by Lade in his correspondence to minimize the significance of the data. But it appeared likely that newspaper readers would gain quite a different impression.

It later developed that segments of the Health Physics Society were also displeased by the publicity. In a letter to the society's board members a month after the annual meeting, R. E. Alexander, chairman of the society's public relations committee, stated that some board members had complained that "the publicity about the paper of E. J. Sternglass . . .

was damaging to the nuclear industry." After defining the "basic publicity objective" of the society, namely, "to let the public know that due to a frankly acknowledged need, we have a new technology, health physics, which will permit them to enjoy the benefits of nuclear energy safely," Alexander went on to say that "while we try to avoid publicizing papers that do not contribute to our basic objective, *there is no way to prevent such publicity absolutely.*" (Emphasis added.)

7

Death before Birth

A FEW MONTHS after the Denver meeting, a letter arrived from *Science* flatly rejecting the Albany-Troy paper. There were two reviews enclosed, both negative. One contained references to nonpublic data available only to the New York Health Department, indicating that the reviewer was a member of that agency. Apparently unaware of the irony, this reviewer claimed that "the number of cases in the Albany-Troy area used in this study are too few to warrant the conclusion." This was, of course, the same number of cases that Lade had used in his 1964 letter to *Science* to prove there had been *no* effect from the fallout, and his conclusion had then been deemed suitable for publication by the editor. And this very same data had also been cited frequently to the same end by the AEC and the New York Health Department. Apparently these figures were only adequate to prove there was no effect.

I now became more determined than ever to pursue the effort to have the full evidence on the worldwide effects of nuclear fallout on childhood leukemia exposed to the scientific community and the public at large. The issue was far greater than the rejection of just another scientific paper. The nature of modern science depends upon the free communication of even the most disturbing ideas, since only through their widest possible examination can the essential process of the gradual correction of errors be accomplished.

And so I began the task of answering each objection raised by the reviewers. One of the points was that the increase in childhood leukemia could probably be explained simply by the increase in the number of children born in Albany-Troy-Schenectady during the postwar years. It was really quite evident that this factor could not possibly begin to explain the fourfold rise in the annual number of cases in an area where the entire population had only increased by 10.5 percent from 1950 to 1960. But now this had to be proved in detail.

I began the laborious task of going through the volumes of the U.S. Vital Statistics, extracting the figures for annual births in the three counties of upstate New York for which Lade had given the leukemia rates. While copying down the numbers, I noticed that there was also a listing for fetal deaths—stillbirths, miscarriages, and other forms of death before birth. Remembering that an increase in fetal deaths had been looked for by J. V. Neel in his study of genetic effects on the survivors of Hiroshima and Nagasaki, I decided to note down these figures as well.

As expected, the gradual increase in annual births during the period from World War II to 1953 was far too small to account for the subsequent increase in the number of leukemia cases. The births had increased by only some 50 percent, while the leukemia cases had risen by over 300 percent.

But the figures for fetal deaths showed something quite unexpected. After steadily declining from 23.8 cases per thousand births in 1941 to a low of 14.4 in 1952, the number of fetal deaths suddenly refused to decline any further. It had only declined to 14.2 seven years later in 1960. This type of change is termed a "leveling off" in the rate of decline. In the case of fetal mortality, such a leveling off was most significant, for the original pattern of steady decline was the result of steady improvements in prenatal medical care which should have enabled the decline to continue until reaching an "irreducible minimum." This minimum had clearly not yet been reached, as was shown by the fact that the decline started again after each discontinuation of testing. More significantly, it reached an all-time low of 11.7 in 1965, two years after large-scale atmospheric testing had ended.

Could it be that this effect on fetal mortality had also been missed by Dr. Neel because his study had not taken into account the effects of fallout on the control population? Was it some kind of statistical illusion, or was it real, and therefore far more serious than even the rise in leukemia?

The library closed for the day, cutting short my investigations. But the implications were staggering. For it was widely known among statisticians that the number of fetal deaths listed in the U.S. Vital Statistics was perhaps no more than one-tenth of the total that actually occurred, since many cases, especially in the early months of pregnancy, were never reported to the authorities. Before the fallout, there had been only two or three leukemia cases per year in the Troy area, as compared with some 150 to 200 reported fetal deaths. The actual number of fetal deaths in the area would probably have ranged from 1500 to 2000. If the relative increase in fetal deaths in Albany-Troy-Schenectady after the fallout was comparable to that for leukemia, then it would involve the loss of hundreds of times as many unborn children as might die from leukemia.

If it was true that fallout caused an increase in fetal mortality, then a similar effect should have occurred not only in the Troy area but also in the rest of New York State, parts of which had been exposed to various other rainouts throughout the 1950s. The next day I returned to the library and went through the data for New York State as a whole. Exactly the same pattern existed: There was a steady decline in fetal deaths toward a minimum of 22.2 per thousand births in 1950, but then the rate actually rose again to 23.3 in 1953 instead of continuing its decline to the 20 per 1000 live births that was to be expected if the normal downward trend had continued. After a brief drop in the mid-fifties, fetal mortality rose once again following the major test series in 1957 and 1958, so that the rate exceeded the expected number by almost 50 percent in 1960. The gap widened steadily as all further improvements in living standards, diet, and maternal health care suddenly failed to have any further effect.

But when I reached 1964, I found the most extraordinary figure of all. In a single year the number of reported fetal deaths in New York State had jumped by 1500 cases. After this it declined once more—the exact same pattern as in the Troy area. In the single year 1964, then, there must have been some ten times this number of reported cases, or, ten to fifteen thousand additional children lost by miscarriage or stillbirth in New York State alone. This tremendous steplike increase for the entire state was clearly connected with the 1961–62 test series, from which large peaks in iodine and other short-lived radioactivity resulted when the spring rains came down in 1963. Unlike the local rainouts of the 1950s, the fallout from these extremely large tests came down much more uniformly over large areas of New York State. Thus there would have been few, if any, unexposed sections, and the state as a whole would show the kind of sharp increases that earlier had been seen only in localities like Albany-Troy.

But could this extraordinary figure be the result of some statistical fluctuation, or a sudden improvement in reporting methods that happened to coincide exactly with the period of the highest fallout levels ever recorded? There was one way to check this very quickly. Unlike the number of fetal deaths, more than 95 percent of all live births are reported to the public health authorities, since nearly all of them take place in hospitals. So if there had really been an increase of some 15,000 fetal deaths in New York State in 1964, then there would have to have been a corresponding sudden drop in the number of children born live the following year. And this is exactly what happened.

For 1962, the total live births in New York State were listed as 354,152. For 1963, the number had increased to 355,760. For 1964, there was a drop to 351,602. But for 1965, there was a sudden decline to 335,628. This was a drop of 15,974 live births, or almost exactly the number of babies lost in 1964 through stillbirth or miscarriage. The rise in fetal deaths must therefore have been real.

It was imperative to make still another test. New York State in the early 1960s was more or less typical of the United States as a whole with respect to the levels of fallout in milk, food, and water. Therefore, the entire country should have shown the same effect: some ten to fifteen times as many fetal deaths in 1964, and a corresponding sharp drop in live births in 1965.

It took only a few minutes to find the figures for the children born live in the United States during these years. For 1964 the number was 4,027,000, and for 1965 it had declined to 3,760,000, a sudden drop of 267,000, the sharpest single decline in the entire history of the United States. And for the entire country the year before, fetal deaths showed a corresponding jump.

It seemed that if there had been about twenty times as many bombs detonated during the 1961–62 test series, there would probably not have been many children born live in 1965.

8

The Crucial Test

WHEN I RETURNED HOME from the library late that night, I was deeply troubled about the implications of these findings, and wondered what course of action I should follow. Should I stop my efforts to publish the article on leukemia in Albany-Troy and concentrate all my energies on this apparently much greater effect of fallout? How could still more convincing evidence best be found? How could scientists and government officials be quickly informed of this discovery so that independent studies to check the findings could be carried out by others as soon as possible?

At that very moment, the French were continuing their atomic tests in the Pacific, while the Chinese were starting to detonate larger-than-ever bombs whose fallout was already drifting over the U.S. If the rise in fetal deaths had truly been caused by fallout, then a simple calculation showed that for each additional megaton of nuclear energy released some 2000 to 4000 infants would be stillborn in the U.S. within a year, and perhaps ten times as many all over the world. And this figure did not even include the many infants who would be born with congenital defects or who would die of cancer and leukemia in the first years of life.

Yet what if my interpretation of the data was wrong? Should I spend many more months or even years gathering more and more detailed data on many additional populations, examining every conceivable alternative explanation for these effects, and only then publish the findings? Clearly this was not just another scientific study, to be handled with more deliberation than urgency. Here, every additional megaton bomb that the French or Chinese tested might mean the loss of thousands of babies, while every additional nuclear weapons system built would increase the certainty that human life would end if these weapons were ever used. The AEC was in the midst of plans to set off another large cratering explosion in Nevada within the next three months in an effort to prove the feasibility of excavating a new Panama Canal by means of hydrogen bombs. Such a test was bound to release large amounts of radioactive debris that would drift all over the United States and northern Europe.

On the other hand, many scientists would surely consider it irresponsible and alarmist to voice concern to the public before all the evidence had been gathered and submitted to the scientific community for detailed consideration. And there would of course be strenuous opposition from all the proponents of nuclear energy for military and peacetime use. Yet by now it was amply clear that I could not expect quick publication of these findings in the widely read scientific journals, such as *Science*. At best, it might take a year to gather the necessary support and definitively ovecome the objections of reviewers.

Ultimately I resolved to take the findings to the public if discussions with colleagues failed to reveal any alternative interpretations of the evidence. Meanwhile, I continued to work at the library, trying to confirm these incredible findings.

One test would be that in a state comparable to New York in economic level and quality of medical care, but where there had been less-intense fallout, the change in the infant mortality decline should have been correspondingly smaller. California, upwind from the Nevada test site, met these criteria, and so I plotted the data for that state. The fetal mortality rate did in fact continue to decline at an undiminished rate during the early 1950s—while New York State had already begun to show sharp rises—reaching rates almost half those in New York. Only in 1955 to 1958, after the large hydrogen bomb tests in the Pacific, did California also begin to show a noticeable leveling off in the rate of decline. But at no time did this leveling show the sharp rises observed in New York State that would be expected from the much more intense short-lived radioactivity produced by the relatively small "tactical" weapons tests carried out in Nevada.

But the most crucial test involved the deaths of infants. For some of my colleagues in the field of public health had pointed out that fetal mortality figures, being incompletely reported, are not nearly as reliable as those for infants who are born live but die before the age of one. And when I examined the infant mortality figures for New York City and New York State, I found that they did in fact show the same peaking in 1964 and 1965 as did the fetal deaths. Especially sharp was the increase in deaths among the infants under 28 days old, which are known to reflect most strongly any effects that occurred during embryonic life.

In order to tie these upward changes in fetal and infant mortality to fallout, it would be necessary to make a detailed comparison between the actual measured levels of radioactivity in the food and milk with the mortality rates to see if the rises and falls coincided. Fortunately, data on fallout levels had been made publicly available beginning in 1957 after a struggle between the Public Health Service and the AEC, which wanted to keep them classified. The first area to check would be New York, for which the fetal and infant death rates were already collected. At the outset, I did not know for certain which of the isotopes in the fallout were causing the principal damage. But then I saw that each time the levels of the short-lived isotopes, such as iodine 131 and strontium 89, shot up to their highest peaks, there was a sharp rise in fetal mortality within a year. The first of these sharp general rises occurred after the very large Nevada, Pacific, and Siberian tests in 1957–58, and the second and highest took place following the tests in 1961–62, the high levels of short-lived isotopes in milk peaking in the spring and early summer of 1963.

Thus it appeared that, in the case of infant and fetal mortality, the short-lived isotopes produced especially by the smaller fission bombs were dominant. These isotopes gave off their radioactivity anywhere from ten to a hundred times faster than the long-lived strontium 90 and cesium 137, or than the carbon 14 and tritium produced by the hydrogen bombs. Inside the body, then, they would give the fetus the highest dose in the shortest time. Certainly iodine 131 could be a major source of damage. In a paper just published in January 1968, Merrill Eisenbud, who had been head of the AEC's New York Health and Safety Laboratory at the time of the Troy incident, reported an actual measurement of the iodine in the thyroid of a 12-week-old fetus aborted in New York City in 1962, the peak year of testing. The fetus had received a thyroid dose ten times as large as that being received by the average newborn infant during the same period. Eisenbud gave this peak average thyroid dose for newborns in the city in 1962 as 200 millirads. So the thyroid of the fetus must have been receiving around 2000 millirads, or 2 rads, a truly enormous dose for this crucial organ compared to the 75 millirads it would normally have received from natural background radiation. And this was just the dose to a single organ from a single isotope. The total effect on the developing fetus from the combined concentration of different isotopes in different organs must have been vastly greater. In this connection, Eisenbud also gave monthly figures on strontium 90 in milk for New York City, and these too showed sharp peaks in 1958, 1962, 1963, and 1964, corresponding to the years of greatest increases in fetal deaths.

But what would be the effects of the long-lived strontium on fetal and infant mortality? The intense radioactivity of the short-lived isotopes generally died out within six months to a year, but the strontium, with its half-life of 28 years, would persist in the environment and in the diet, continuing to build up in the bones and genetic material of the exposed people until reaching an equilibrium level where natural metabolic processes would remove it at the same rate as it was being taken in. Studies had shown that this peak level was generally reached some four to five years after continuous intake began.

Therefore, it seemed likely that any effects of strontium on infant mortality would parallel this accumulation in the bodies of the parents. If this was so, then after the initial sharp rise caused by the short-lived isotopes there would be a dip, followed by a gradual rise culminating in a second, broader, lower peak extending generally between the third to fifth years. Thereafter, if no additional strontium was added to the diet, there would be a slow decline that would probably accelerate rapidly after a few years as the strontium in the environment was dissipated and diluted by natural processes.

When I discussed the findings with Dr. Barry Commoner, now at Washington University in St. Louis, he suggested that I compare the leveling of the decline in fetal and infant mortality with the measured amounts of strontium 90 on the ground and in the milk for different areas of the United States. It took only a few days to discover that the pattern followed closely the levels of strontium 90 that accumulated in the environment after the onset of hydrogen bomb testing in the early 1950s. Furthermore, the graphs consistently showed two peaks in tandem—a sharp peak within a year after each test series, when the levels of short-lived isotopes as well as strontium shot up, followed by a second slower rise culminating between three and five years later. The second peaks were especially high, probably because each of the enormous fusion bombs had actually produced hundreds of times as much strontium 90 as one of the earlier atomic fission bombs, even though the hydrogen bombs had been advertised as being much "cleaner." For, as Ralph Lapp and the British physicist Joseph Rotblat had each discovered independently in 1954, in order to get a "bigger bang for a buck" as U.S. Secretary of Defense Charles Wilson put it, Edward Teller and his weapons engineers had surrounded the hydrogen bombs with cheap, abundant uranium 238. As a result, the total explosive force could be doubled at no additional cost, but the levels of strontium 90 in the bones of living creatures were vastly increased.

The final task that remained was to make certain that there existed no other known explanation for the halt in the decline of infant mortality in the United States. Various of my associates in the University of Pittsburgh School of Public Health had said that to the best of their knowledge no other cause had been found. It was suggested that I review the results of an international conference on the problem, held in Washington, D.C., in May of 1965.

The summary of the conference revealed that extensive studies had been made comparing the U.S. with five European countries where there had been much less of a slowing in the infant mortality decline, or none at all. These studies had failed to find any explanation for the sudden worsening of the situation in this country. However, fallout had not even been considered as a possible factor.

According to the report, the U.S. infant mortality problem had become so serious since it was first noticed in 1960 that comparative studies had been undertaken in Scotland, England, Wales, Norway, Denmark, and the Netherlands. The 1965 conference brought together the investigators from each of these countries in an effort to determine "the reasons for the position of the infant mortality rate in the United States."

As the introduction to the report put it:

Two features were noted about the infant mortality rate in the United States in the 1950s: (1) the virtual halt in the rate of decline of infant mortality after a long period of rapid decline, and (2) the unfavorable position of the United States infant mortality compared with that of many other countries. These two points were viewed with concern because health authorities had for many years pointed with pride to the high rate of decrease in infant mortality in the United States.

In his opening remarks, Dr. I. M. Moriyama, chief of the Office of Health Statistics Analysis of the U.S. National Center for Health Statistics, posed the problem as follows:

One of the intriguing questions is why the rate of decline of infant mortality rates for countries such as the United States, England and Wales, and Norway has been checked, whereas the rates for Denmark and the Netherlands have continued to decline without apparent interruption.

But what was more surprising was what Dr. Moriyama added:

> Incidentally, the change in mortality conditions was not peculiar to the period of infancy. The rates at other ages also leveled off in the United States, but these changes came several years after the beginning of the deceleration in the infant mortality rate.

This delay in effects on the older population was, of course, consistent with the characteristic established pattern of delay in appearance of radiation-induced cancer and leukemia. And I had found that fetal and infant mortality changed within a year after the exposure took place.

An observation made by Dr. Samuel Shapiro, director of research and statistics of the Health Insurance Plan of Greater New York, seemed to lend further support to the fallout hypothesis. Dr. Shapiro remarked that the birthweight of U.S. infants had mysteriously declined since the early 1950s. This was significant, since a decline in birthweight had been well established as one of the effects of radiation on unborn children in the studies of the Hiroshima and Nagasaki survivors. And extensive laboratory animal studies had long shown that stunting of growth and reduced birthweight were produced by irradiation of the fetus, while similar effects had been observed among human infants accidentally exposed to X-rays prior to birth. Furthermore, a slight reduction in birthweight, such as sometimes appears in infants whose mothers had German measles during pregnancy, was known to greatly increase the likelihood that the infants would die in the first year of life. As it was put later in the report of the conference: "Low-birthweight infants have a much greater chance of dying, and hence contribute significantly to neonatal mortality."

Dr. Shapiro also noted that "the decrease in the rate of decline was more serious in the nonwhite population." The large series of bomb tests that began in Nevada in 1951 would logically have had the greatest effects on the large black populations of the Southern Atlantic and Gulf states. These included the states of heaviest rainfall, over which the fallout from Nevada was blown by the prevailing eastward high-altitude winds, spreading radioactive rain on the crops that formed the staple diet of the sharecroppers and farm laborers of the South. And their children, with the poorer diet, poorer sanitary conditions, and poorer medical care prevalent among the nonwhite populations, would, if they were born slightly weaker and less resistant than normal, have a greatly reduced chance of survival as compared with white children, who are generally far better off in all three respects.

According to the report, it appeared that infant mortality patterns in European countries also fitted the fallout hypothesis. According to Charlotte A. Douglas, a public-health physician from Edinburgh, "In Scotland, there had been impressive declines in maternal, fetal, and infant mortality from 1935 until the early 1950s, when the decline in mortality rates became more gradual. Since then there has been only slight improvement." Thus there was a close coincidence in the times when this leveling trend began in Scotland and the U.S. Similarly, in England and Wales, as reported by Dr. Katherine M. Hirst: "The total infant mortality has declined for years, but it began to level off during the 1950s, as did that of the United States." And again, just as in the U.S., there was a large increase in the number of low-birthweight babies. The same story emerged from the account of Dr. Julie E. Backer for Norway, where the clouds drifting across the North Atlantic would have rained down their fallout most heavily on the coastal mountains rising from the sea.

However, the reports on Denmark and Holland were quite different. In these countries there had been little or no leveling off in the rate of decline. A pattern was beginning to emerge. It was the more northerly countries with the heaviest rainfall that were showing the greatest effects on infant mortality. Even within England, there was a reduction in infant and fetal death rates going southward.

The high-altitude fallout clouds from the Nevada test site, carried across the Atlantic in a northeasterly direction by the prevailing jet-stream wind currents, would have deposited their radioactivity on the northern parts of Europe with much greater intensity than on the southern parts. This pattern was confirmed by the lack of any significant halt in the steady decline in infant mortality in France throughout the early period of Nevada testing. There, the first leveling of the decline did not occur until after the first tests in Algeria in 1960 and the large hydrogen bomb test series in the Pacific in 1961–62. These tests resulted in substantial fallout in France. And the same situation existed in Canada, which had been largely untouched by the fresh fallout from Nevada as it was blown northeastward across the U.S. and on to northern Europe.

An even more striking confirmation of the thesis that these geographical differences in mortality rates were related to fallout was contained in a study by Dr. Bernard Greenberg, a biostatistician at the University of North Carolina, whose work was summarized in the report. Since Sweden had continued to show a much smaller percentage of low-birthweight babies than the U.S., Dr. Greenberg undertook to find out if there was some cultural or genetic difference among people of Scandinavian stock that made their children less susceptible to this condition. Taking the two states of Minnesota and North Dakota, Dr. Greenberg compared the number of low-birthweight babies in counties containing the highest percentage of Scandinavians with the number in counties containing very few Scandinavians. He did indeed find a small difference of about 10 percent, the counties with more Scandinavians showing the smaller number of low-birthweight children. But this percentage was far too small to explain the much larger difference between the U.S. and Sweden. However, he also found something he could not explain at all, namely: *All* the counties in North Dakota, regardless of the percentage of Scandinavians, showed a much higher incidence of low-birthweight babies than did the counties in Minnesota.

This was odd. But then I remembered an article I had read many years before about fallout in North Dakota. It was entitled "The Mandan Milk Mystery." In the back files of *Scientist and Citizen* (now called *Environment*), a magazine then published by the St. Louis Citizens' Group for Nuclear Information, I located the article, written by E. W. Pfeiffer. It described

how the Mandan, North Dakota, milk sampling station operated by the AEC had, for some unknown reason, consistently
shown the highest concentration of strontium 90 in milk among
all the states where such measurements were taken. In fact,
in May of 1963, Mandan had shown the highest levels ever
recorded anywhere in the United States. The levels were consistently much higher than in nearby states, notably Minnesota.

This, then, was a direct correlation between the amount
of strontium 90 in milk and the incidence of low-birthweight
babies. It should therefore be possible to find large geographical
differences in infant mortality based on differences in fallout
levels. The various regions should show rises beginning at different times. These rises should come first in the states in the
path of the fallout from the early tests in New Mexico and
Nevada—the southeastern states of heavy rainfall, along the
Gulf of Mexico and the Atlantic coast from Forida to about
New Jersey. These were the regions over which the high-altitude
clouds generally drift from west to east, carried along by the
"jet streams," the constant 60- to 120-mile-an-hour winds blowing northeastward at altitudes of 25,000 to 40,000 feet. Only
later, after the hydrogen bomb testing began in the Pacific in
1953, should the high-rainfall states in the northern U.S. begin
to show an upward trend. For in these tests, which continued
up until the test-ban treaty of 1963, the radioactivity was deposited high in the stratosphere, beyond the influence of the prevailing winds, and so it sifted down much more uniformly around
the world. But since this high-altitude debris was also eventually
brought down mainly by rain and snow, then there should
still be clear differences in infant mortality between the wet
and dry regions of the U.S.

Searching among the reports published by the U.S. National
Center for Health Statistics, I located one that had graphs of
infant mortality for every state from 1935 to 1964. And indeed,
the southeastern states along the Gulf and Atlantic coasts
showed the first sharp leveling during the 1946–50 period following the first test in New Mexico, while the states farther
to the north did not show this trend until later in the 1950s
when the tests were moved northward to Nevada. And overall,

from 1935 to 1964, the least change in the steady decline occurred in the low-rainfall states of the Southwest. In New Mexico, for example, except for a brief halt in the downward trend following the first and only atmospheric test carried out in this state in 1945, the pattern was essentially the same as for France. The decline resumed and showed only a slight degree of slowing until the early 1960s, when the large amount of debris from the South Pacific tests began to raise the levels of radioactivity everywhere.

Furthermore, the lower-income groups—mainly the nonwhite population—showed the greater change, so that the effect first became apparent in this group. In fact, in some states, the decline among this group did not just slow or stop. In Arkansas, for instance, the nonwhite infant mortality rate actually began to climb again from a low point of 30 per 1000 live births in 1946 to 39 per 1000 in 1949, gradually rising still further to a high of about 42 after the peak of testing had resulted in the highest levels of radioactivity ever in the diet in 1963. In the face of generally improving living standards, such a trend was very hard to understand any other way.

In sharp contrast, the nonwhite infant mortality rates for dry Arizona and New Mexico kept right on declining. Yet the medical care and general living standards of the Indian populations in these states was not significantly better than for the black population of Arkansas. In New Mexico, the rate for nonwhites in 1952 was close to 100 deaths per 1000 live births. During the ensuing decade of heavy testing, the rate declined continuously to 40 by 1961, just as steady a rate of decline as that for the white population with its much higher living standard.

But the most convincing evidence that it had to be fallout rather than ordinary chemical pollution or any genetic, cultural, medical, or economic factors came from the evidence for Hawaii. Here was an area that originally had roughly the same high infant mortality rate among nonwhites as the southwestern states of the continental United States: about 80 per 1000 births in the early 1930s. It showed a sharp downward trend in the late 1930s that brought the rate to about 28 in 1945. But then this decline suddenly halted, and the rate actually rose during 1946–48, shortly after the Hiroshima and Nagasaki detonations and the first tests at Bikini and Eniwetok, all located directly upwind from Hawaii. The rates for both whites and nonwhites afterward resumed their decline, reaching about 21 by 1951–52. Shortly thereafter, however, they once again not only stopped their decline but actually rose to another peak of 25 between 1957 and 1960, following the large hydrogen bomb tests conducted during 1953–58 in the Pacific. Not until four years after the end of large-scale atmospheric testing in 1962 did Hawaii resume its decline, finally going below 20 in 1966, the lowest rate in its history. The hypothesis that it was the fallout coming down with the rain fitted perfectly: The annual rainfall in Hawaii was among the highest in the world—some 100 to 200 inches as compared to less than ten for New Mexico and Arizona in the U.S. Southwest.

The worst situation in the continental U.S. existed in Mississippi, a state directly in the path of the fallout from many Nevada tests and with poor medical care and an annual rainfall of 49 inches. There, the nonwhite infant mortality rate had shown a promising decline even throughout the period of the Depression, dropping from 82 per 1000 in 1930 to a low of 40 by 1946, quite close to that for the white population. But instead of declining further, the nonwhite rate leveled off in the early 1950s and then actually started to climb sharply. By 1963 it was back up to 58, an absolute increase of 45 percent in the face of a generally rising standard of living and improved diet and medical care. And once again, just as in Hawaii, the rate renewed its previous decline after the cessation of atmospheric testing, reaching 53 by 1966, the last year for which data were available.

Furthermore, examination of the official fallout measurements showed that the high rainfall areas of the south and east did indeed have two to three times as much strontium 90 in their soil as the dry states of the western mountain region. By plotting the figures for twelve typical rural and urban states on a graph, I was able to ascertain that the upward deviation infant mortality was directly related to the amount of strontium 90 deposited.

The total excess infant mortality for the United States as a whole was truly enormous. Using the calculating methods employed by Dr. I. M. Moriyama, it was possible to estimate what the mortality rates would have been if the decline had not been interrupted. In the fifteen years between 1951 and 1966, the total number of infants in the United States who died in the first year of life exceeded the norm established during the previous fifteen years by 375,000. The infants were not dying to any noticeably greater degree of bone cancer and leukemia, the effects well known to be produced by strontium 90. Instead, they were dying a little more frequently of respiratory diseases, infections, and immaturity, conditions that apparently had nothing to do with the kind of gross effects everyone had been led to expect from radiation, and that would be more noticeable among those who had poorer diets and medical care.

Even more staggering were the figures indicating that for every infant who died in the first year of life there were five to ten who died prior to birth, so that the excess numbers of fetal deaths, spontaneous abortions, and stillbirths must have reached anywhere from two to three million in the United States alone.

In addition, after 1955 there had also been a sudden slowing down in the steady decline of maternal mortality. Between 1937 and 1955 the average rate of decline in the number of women dying from complications of pregnancy and childbirth had been 12.8 percent per year. This rate of decline slowed down drastically to 2.1 percent per year between 1956 and 1962. Even more disturbing, instead of a further decline, there was actually a rise in the number of maternal deaths between 1962 and 1963, the year when fallout reached the highest levels ever recorded and fetal deaths began their sharp rise. Thus in 1963, a total of 1466 women died from complications of pregnancy and childbirth in the United States. The calculations indicated that if the previous downward trend had continued instead of leveling off, close to a thousand of these mothers would not have died in that one year alone. For the whole world combined, the figure would have been ten times as large.

It did not take long to discover that beginning in 1966, some three years after the test-ban treaty was signed, when only the French and Chinese continued testing in the atmosphere, the infant and fetal mortality rates all over the U.S. and Europe suddenly and quite unexpectedly began to decline once more. According to the latest Monthly Report of the U.S. Office of Vital Statistics, the drop in infant mortality that began for the U.S. after 1964, when it had reached a peak of 24.8 per 1000 births, had indeed continued into 1968, reaching 21.7, and it showed no sign of halting. Even Sweden had resumed its decline at the same rapid rate that it had shown prior to the onset of testing. And there, the number of deaths per year had already been so far below that of the U.S. that any further decline seemed extremely unlikely. Here was clear proof that the "irreducible minimum" in infant mortality had not been reached either in Sweden or the United States.

By 1964, the year before the international conference on infant mortality had been held, seventeen other nations in the world had reached a lower level of infant mortality than the United States, according to a report published by Dr. Helen C. Chase of the U.S. Office of Health Statistics Analysis in 1967. These included the Netherlands, 14.8; Norway, 16.4; Finland, 17.0; Iceland, 17.7; Denmark, 18.7; Switzerland, 19.0; New Zealand, 19.1; Australia, 19.1; England and Wales, 19.9; Japan, 20.4; Czechoslovakia, 21.2; Ukraine, 22.0; France, 23.3; Taiwan, 23.9; Scotland, 24.0; and Canada, 24.7. And for the United States in that year the figure had been 24.8, or 68 percent higher than for the Netherlands, when as recently as 1947 the United States actually had a lower infant mortality than that country. Among all these nations, it was the U.S. which was exposed to the most intense fallout from the tests in Nevada and the South Pacific.

Most of these countries had been exposed to other environmental pollutants to the same or even a greater degree than the U.S. Tobacco, food additives, pesticides, and drugs were used throughout the world since World War II. Neither could ordinary air pollution be the principal cause, since unpolluted rural states such as Hawaii and Arkansas showed far greater upward deviations in infant mortality than any of the heavily industrialized urban states in the northeast.

The conclusion was inescapable. There seemed to be no other single factor that could account for such sudden and dramatic changes on a worldwide scale. Only radioactive fallout acting mainly on the early embryo could explain these facts.

After discussing any findings with colleagues in the Federation of American Scientists, it was agreed that they would be made public at a meeting of the Pittsburgh F.A.S. chapter on October 23, 1968. The day before the meeting, I submitted two copies of the report to *Science*. Because of his interest in the subject, I also sent a copy not intended for publication to Eugene Rabinowitch, editor of the *Bulletin of the Atomic Scientists*, a journal that had long been concerned with the possible biological effects of fallout.

After delivering my paper at the F.A.S. meeting on the morning of the twenty-third, I was interviewed by Stuart Brown, a reporter from the Pittsburgh television station KDKA. A few hours after the conclusion of the interview, the phone at my office rang. It was Stuart Brown, who said that he had just called Philip Abelson, the editor of *Science*. Brown thought I ought to know what Abelson had said.

According to Brown, Abelson had told him that my paper had been rejected by an "independent committee." This was a most unusual statement for the editor of a scientific journal to make to the press. According to long-established tenets of professional ethics, such journals are supposed to keep all editorial correspondence completely confidential. Furthermore, the paper I had just presented had been mailed to *Science* only the night before and thus could not possibly have been reviewed by an "independent committee." Abelson had obviously been referring to the revision of the Troy paper on the rise in leukemia, which contained none of the new data on fetal and infant mortality.

Next, Brown said, Abelson had gone to the files, pulled out my folder, and read statements from the supposedly confidential report of one of the reviewers that my paper was "weak in its scientific methods" and its findings were "sweeping and sensational." And at the end of the conversation, Abelson advised Brown against using any of my findings on the air.

9

Both Young and Old

MEANWHILE, I continued my search, for the investigation of infant mortality had revealed still another possible dimension of the effects from fallout. As Dr. I. M. Moriyama put it in the introduction to his 1964 report entitled "The Change in Mortality Trend in the United States":

> The same kind of change in trend observed for infants appears to be taking place in the death rates for other ages.

Regarding the overall trend, he added:

> The failure to experience a decline in mortality during this period is unexpected in view of the intensified attack on medical problems in the postwar years . . . there has been a growth in the volume and scope of health services in prevention, diagnosis, medical and surgical therapy, and rehabilitation, and also an improvement in their quality. The rapid growth of health insurance plans has made high quality medical care readily accessible to ever-increasing numbers of people. The rising level of living has resulted in improvement of work and home environment, quality and variety of food, educational attainment, and facilities

of recreation. Developments in medicine arising from the exigencies of a global war have become readily available for application to civilian health problems. At no time in the history of the country have conditions appeared so favorable for health progress.

In this setting, it would seem reasonable to expect further reductions in mortality. On the other hand, the possible adverse effects on mortality of radioactive fallout, air pollution, and other manmade hazards cannot be completely ignored.

In the School of Public Health library, page after page of data showed the same dramatic upward changes for chronic diseases as had occurred for childhood leukemia and fetal and infant mortality. Deaths from all types of noninfectious respiratory diseases such as lung cancer, emphysema, and bronchitis had increased especially dramatically among all age groups, as had deaths from certain other types of cancer. The overall life expectancy, particularly for adult males, had begun to level off and then actually declined again after decades of steady rise in this country and in northern Europe.

Most significant of all were the death rates due to all types of childhood cancers in the U.S. For young people of both sexes, white and nonwhite alike, there were sudden, steplike jumps in the cancer rate between 1948 and 1951. For the white children, the rate doubled during these three years. The rate for the nonwhite children, which had held basically steady between 1930 and 1948, tripled during the same three-year period. This was to be expected if fallout was the cause, since by far the largest portion of the nation's nonwhite population lived in the regions where the fallout from the early tests came down.

When had this overall trend begun? Dr. Moriyama's report indicated that deaths from respiratory diseases and childhood cancers either declined steadily or held level throughout the 1930s and most of the 1940s, the period of rising air pollution and the tripling of cigarette consumption. Between about 1948 and 1950, three to five years after the detonation of the first bombs in New Mexico and Japan and the onset of atomic testing in the Pacific, the death rates from these diseases suddenly began to shoot up. For example, the annual death rate among white males seventy-five to eighty-four years old from respiratory diseases (not including influenza and pneumonia) was close to 110 per 100,000 per year in 1934. By 1948 it had declined to an all-time low of about 70. But after this it shot up to 190 by 1960.

The sharp upward changes in the rates for these types of chronic diseases were reflected in an overall leveling in the decline of death rates for the United States as a whole. But this effect was particularly serious in certain states, where the death rates actually rose again after decades of steady decline. As Moriyama pointed out: "In twelve states and the District of Columbia there appears to be a marked rise in the crude death rate during the past five to ten years as represented by the trend for North Carolina." The list of the states showed that they were all either southern states to the east of New Mexico and Nevada, or states directly to the northeast of the Nevada test site. The states were Alabama, Arkansas, the District of Columbia, Louisiana, Missouri, North Carolina, South Carolina, Tennessee, and West Virginia in the South, and Nevada, Colorado, South Dakota, and Wyoming in the West. None of these states was heavily industrialized or noted for its air pollution. In fact, the opposite was true.

The number of excess deaths in the U.S. resulting from these upward changes in the death rates was calculated by Dr. Moriyama as being 300,000 during 1956–60. This was the period of heavy Nevada testing. The fallout and the excess deaths continued as testing resumed in 1961. According to Moriyama, "The estimated excess deaths is about 85,000 deaths for 1961, and 131,000 deaths for 1962." The "excess deaths" had jumped from an average of 60,000 per year during the earlier period of testing to 85,000 in 1961, the year that the Russians detonated the largest megaton weapons ever exploded. In the following year, when nearly 70 megatons of fission energy were detonated by the U.S. and the Soviet Union—the highest megatonnage ever exploded in a single year—the excess deaths in the United States alone reached 131,000.

The same trends were also evident in other parts of the world. A detailed report on changes in mortality trends for

England and Wales by Hubert Campbell at the Welsh National School of Medicine showed that in those countries the mortality rates for the very young and the very old followed exactly the same pattern as in the U.S. Beginning about 1953–55 the total death rates for 1- to 4-year-olds suddenly refused to decline further. For the 5- to 9-year age groups, which the studies of Stewart and MacMahon had shown to reflect most strongly the cancer-causing effect of irradiation during early development, the mortality rate actually started to climb again. Beginning the year after the first Nevada tests, there was also the same sudden halt in the rapid decline of maternal mortality associated with complications of pregnancy and childbirth as in the U.S. These rates actually turned sharply upward in 1960–61 for the youngest group of women (15 to 24 years of age) for the first time in modern history. And there was the same sharp rise in cancer deaths of all types for the age group 5 to 14 years, following some three to five years after the New Mexico test and the detonations in Japan in 1945. Both male and female death rates jumped in a steplike fashion: from about 30 per million per year to about 60 for boys, and from 25 to about 50 for girls, all between 1948 and 1951, exactly as in the United States. That these rises could not be due to changes in statistical or classificational methods was emphasized by Campbell: "There has been no important change in the classification of these diseases during this period. . . ."

For all age groups in England and Wales, Campbell's report showed tremendous rises in leukemia for both men and women, beginning suddenly between 1947 and 1951, with the sharpest changes for the very young and the very old. Thus, whereas the leukemia death rate had remained fairly steady for men 75 to 84 years old during the fifteen years between 1931 and 1946, ranging from 50 to 80 deaths per million individuals each year, by 1954 the rate had increased to about 200. It reached 350 by 1959, an increase of about 500 percent. During the same period, the leukemia rate remained unchanged for the middle-aged group 45 to 54 years old, but it increased some 50 percent for boys 5 to 14 years of age. This was the rise that had prompted Dr. Stewart's study.

The data for Japan, prepared by a group of public health physicians and statisticians from the Japanese Institute of Public Health, was particularly significant, since Japan was not only exposed to the fallout from the Hiroshima and Nagasaki bombs, but also received the radioactive debris from the U.S. Pacific and Soviet Siberian tests.

The report showed that three to five years after the fallout from Hiroshima and Nagasaki descended in 1945 the cancer rate for the 10- to 14-year-old children all over Japan tripled from 10 to 30 cases per million population, gradually climbing further to 40 cases by 1955 and to 50 by 1963, a fivefold increase during the period of heavy testing. For the youngest children zero to 4 years old, the increase was less, once again confirming the hypothesis that radiation was the causative factor as in the case of Troy. Again, the rates for the middle-aged group remained level, while the rate for those over 80 went up as elsewhere, in the case of Japan from about 3000 to 8000 per year per million individuals.

Here then was the confirmation of why the studies of the Hiroshima and Nagasaki survivors had not revealed any effects on their children. Everywhere in Japan, mortality rates had gone up due to the fallout, so that there was little or no difference between those survivors exposed to the direct flash and those who received the fallout in their diet over the years that followed.

And after the large hydrogen bomb tests, deaths due to noninfectious lung diseases such as emphysema and bronchitis suddenly stopped declining in Japan after 1955. This was the year after the Pacific and Siberian tests filled the air all over the world with radioactivity. In the next two years, deaths due to bronchitis, which had been dropping rapidly from 150 per million population in 1950 to a low of 40 by 1955, actually began to rise again. Thus, in the period from 1945 to 1955, when industrial growth and the accompanying smog and chemical pollution had been very great, these respiratory diseases had been declining. As in the United States, they rose only after the enormous increase of atmospheric radioactivity.

If the major factor was fallout and not the pollution produced by industry and the automobile, then Chile provided an excellent chance to test this hypothesis in a country of low industrialization, as it was the only South American country for which detailed mortality-trend data was available. Since Chile was located on the west coast of South America, facing the prevailing winds from the South Pacific that release their moisture on the steeply rising slopes of the Andes, there should be upward changes in mortality following the first two series of Pacific A-bomb tests in 1946 and 1948. And these increases should be even more noticeable for the heavy series of tests beginning in 1952, involving the "dirty," uranium-clad hydrogen bombs that had produced such massive amounts of fission products.

The Chile mortality graphs instantly confirmed this prediction, especially the plot of mortality for the infants dying between the ages of one month and one year, which showed an initial rise between 1947 and 1949 after the first Bikini and Eniwetok tests. Far more serious was the sudden and complete reversal of the overall infant mortality trend, from a steady decline to a continuous rise beginning in 1954 and persisting until 1960, the last year for which data were available.

And the same change had taken place in the total mortality rate for all ages combined. There had been a steady decline after 1933, except for small rises during the second half of the 1940s, but then between 1953 and 1955 there was a sudden and complete end to the decline for both men and women, continuing for as long as the data had been plotted.

In the words of the report's authors, a group of Chilean public health specialists:

> The significance of this trend is evident if the mortality for 1960 is estimated on the basis of regression for 1933–53. The expected rate was 8.6 and the observed rate was 12.3, which means that 28,024 of the total 93,265 deaths registered in 1960 would not have taken place if the previously described trend had continued.

This did not mean that cigarettes or air pollution were not significant factors in chronic lung disease, or that heavy metals, pesticides, food additives, and other pollutants were not adversely affecting worldwide health. The phenomenon of synergism, in which combinations of two or more biological agents have a much greater effect than one alone, is well known to modern science. For instance, it has long been known that uranium miners have ten times the normal rate of lung cancers because of their breathing of radioactive gas in the mines. But those who smoked died of lung cancer at one hundred times the normal rate.

However, statistics from all over the world kept indicating that radiation was the dominant factor in these worldwide changes of mortality trends. It made no difference what the social or economic system was, nor how much medical care was available, as in the very different cases of the Netherlands versus Chile. It made no difference whether infant mortality was high or low to begin with, as in Mississippi versus Sweden. It did not matter whether there was any air pollution, or what the genetic, cultural, or dietary differences were. There was only one way to explain these worldwide, synchronous, and totally unexpected changes that did not stop at any national boundaries nor at the edges of the seas. Only the introduction of some new and enormously powerful biological agent on a worldwide scale could produce such sudden rises in death rates that could almost be termed epidemics. And this new agent clearly seemed to be the fallout that had been released into the atmosphere in quantities equivalent to tens of millions of pounds of radium, the most powerful biological poisons yet created by man, circling the world in a matter of a few weeks and attacking mainly the weakest in every living species—the developing young and the very old.

10

The Clouds of Trinity

THE GEOGRAPHIC PATTERNS of the changes in worldwide leukemia and infant mortality trends between 1945 and 1955 clearly matched the patterns of fallout. But so unbelievable and far-reaching a conclusion required much more evidence before the possibility of any other explanation could be ruled out. Laboratory-animal experiments had shown that various pesticides, drugs, food additives, and heavy metals could apparently cause cancer and congenital defects, while air pollution and mothers' cigarette smoking were believed to be linked to fetal and infant mortality.

There was one test, however, that would effectively rule out these agents as the principal factors in the increase. It would involve the first nuclear explosion ever set off by man. Code-named Trinity, this explosion took place at dawn on July 16, 1945, in Alamogordo, New Mexico. Before this, there was no nuclear fallout in the environment, so this explosion would have to have produced a clear effect on mortality rates wherever the fallout descended.

But where had the fallout come down? There was at that time no elaborate countrywide network of fallout-measuring stations. However, many eyewitness accounts of this historic explosion had subsequently been published, and, among these, a book called *Day of Trinity* by Lansing Lamont reported the event in greatest detail. Lamont, a *Time* magazine reporter, had gone to considerable lengths to trace the direction of the drifting radioactive debris, studying weather maps and conducting extensive interviews with many of the scientists involved.

According to Lamont, shortly before the final countdown at 5:10 A.M. on the morning of July 16, Dr. Kenneth Bainbridge announced to the scientists at the various observation posts that the winds close to the ground were blowing north, toward where Dr. Robert Wilson was manning an observation post 10,000 yards from ground zero. An instant after the flash of the detonation at exactly 5:29 A.M., the churning fireball detached itself from the ground and shot upward, followed by a column of radioactive dust, penetrating the overcast at 15,000 feet. The column of dust continued upward to an altitude of 40,000 feet, where it spread out in the mushroom shape that was later to become so familiar.

The lower part of the mushroom's stem was blown north toward Wilson's position, which had to be quickly evacuated. Simultaneously, Dr. Luis Alvarez and Navy captain William Parsons, flying high above the cloud cover just to the west of the test site in an observation plane watched the head of the mushroom penetrate the overcast and break up into three distinct sections. These sections drifted off in different directions, generally to the northeast and east. As recounted by Lamont, the largest of the three sections, a dense white mushroom trailed by a dusty-brown streamer, drifted off in a direction just slightly north of east. Meanwhile, the low-altitude fallout from the stem of the mushroom cloud continued north and northeast until it covered an area about 30 miles wide and 100 miles long, gradually settling to the ground in a white mist of intense radioactivity.

By three o-clock in the afternoon, the readings on the radiation counters monitored by Alvin Graves and his wife, Elizabeth, observers assigned to the little town of Carrizozo some 40 miles just slightly north of east from the test site, started to climb rapidly. By 4:20 P.M., eleven hours after the explosion, the counters shot off scale and Alvin Grates called Dr. Stafford Warren, the chief medical officer in charge of radiation monitoring. As Lamont put it, the fate of the little town hung in the balance while the scientists and Army officers decided whether or not to evacuate it. Ultimately, they held off, and within an hour the fallout readings had dropped. Lamont reported that these were difficult hours for Dr. Warren and the officer in charge of the entire project, General Leslie R. Groves. "The medical dangers were most immediate of all," Lamont wrote, "but, in addition, both men knew that the Army was not too eager to pursue too diligently the possibilities of widespread fallout."

From the fact that it had taken the fallout particles some nine to eleven hours to reach Carrizozo, it was possible to determine which portion of the cloud had gone eastward. According to the AEC publication *The Effects of Nuclear Weapons,* typical fallout particles descend at a speed of about 5000 feet per hour, while smaller particles fall more slowly. Allowing about half an hour for the cloud to travel the 40 miles to Carrizozo at the usual speed of the jet-stream air currents, then, it meant that the particles that caused the Graveses' radiation counters to start climbing must have taken about eight to nine hours to descend at a rate of 5000 feet per hour. Therefore, the cloud that passed over Carrizozo must have been at an altitude of between 40,000 and 45,000 feet.

Thus, it had to have been the uppermost section of the mushroom cloud that drifted just slightly north of east. And since the winds became more northerly with decreasing altitude, the lower, smaller sections would have gone increasingly northward. This estimate was confirmed by the fact that the more northerly towns were the first to receive the fallout, and those to the east were the last. The low-altitude fallout, therefore, had come down mostly in New Mexico to the northeast, while the highest portions would have been carried more nearly eastward across Texas, Oklahoma, Arkansas, and the whole southeastern U.S., where they would be brought down mainly by the rains.

On the basis of these estimations, then, any upward changes in infant mortality should be found to some degree in the sparsely populated areas of New Mexico itself, and to a greater extent in the more heavily populated states to the east, north-east, and north. Among the more distant states generally to the east, those with the heaviest rainfall should show the largest upward changes, since the high-altitude cloud carried the small-est particles, which would largely remain aloft unless brought down by rain. Furthermore, the states near the Atlantic sea-board to the northeast, such as North Carolina and Virginia, should be affected less than nearby Arkansas, Louisiana, and Mississippi, for the cloud would have gradually fanned out and the short-lived radioactivity would have steadily diminished in intensity with the passage of time.

The five-year period 1940–45 was the longest period imme-diately preceding the test during which a steady decline in infant mortality had existed in every state in the U.S. Thus, the amount of upward deviation from this rate of decline would provide a measure of any changes occurring after the explosion.

When the infant mortality figures were plotted on a map of the U.S., the pattern began to emerge. The states directly to the west and far to the northeast of New Mexico kept declin-ing at the 1940–45 rate. In fact, in some cases the decline was actually somewhat faster, due to the introduction of sulfa drugs and antibiotics, since the greatest cause of infant deaths was the infectious diseases that these new drugs succeeded in cutting back. But each year after 1945 and beginning strongly in 1947, there was a growing excess infant mortality in the states of the Gulf and Atlantic coasts to the east and northeast of the test site, amounting to as much as 30 to 40 percent. This pattern extended over the entire southeastern part of the country, from Texas, Arkansas, Louisiana, and Mississippi all the way across Alabama and Georgia to South Carolina, North Carolina, and Virginia, and existed both for the poorer non-white as well as for the socio-economically better-off white infants.

Two of my colleagues, Donald Sashin and Ronald Rocchio, became quite concerned about these findings and offered to work out a computer program to calculate and plot the infant mortality rates for every state, thus removing any possible subjective bias. The pattern that emerged from the computer was essentially the same except for one striking difference. For some reason the computer maps showed excesses in infant mortality for the north-central region of the U.S., in Montana, Idaho, Wyoming, and especially North Dakota. In 1946, even before the increased infant mortality manifested itself to the east of New Mexico, North Dakota was showing a 19 percent increase, reaching 32 percent in 1949. But the low-altitude fallout that went northward from the Trinity test could not possibly have been significant in these distant states so directly to the north.

The mystery was solved when a colleague happened to show me a copy of an AEC publication entitled *Meteorology and Atomic Energy.* In the opening chapter, dealing with the history of the atomic energy industry, the report explained that in 1944 the first of a series of giant nuclear reactors had gone into operation in Hanford, Washington, to produce the plutonium for the Trinity bomb. The reactor was located in the dry eastern edge of the state of Washington, directly upwind from Montana, Idaho, and North Dakota. Because the operating engineers did not have sufficient experience with these enormous new reactors being built under wartime pressures, large releases of radioactive gases occurred. As the AEC report described it:

> As soon as a charge of fuel came out of the plutonium production reactors, a large source of gaseous effluent was encountered. For the plutonium produced to be removed from the uranium and other fission products, it was necessary to dissolve the fuel by various chemical reactions. During the early stages of this process, all the noble-gas fission products, notably radioactive isotopes of xenon and krypton, were released. It was not feasible to remove them by a filter system; they were released to the atmosphere in rather large quantities.

Still more disturbing was the statement that "large quantities of radioactive iodine were involved." Additionally, when the reactor's fuel elements would occasionally catch fire, krypton and biologically more hazardous fission products such as stron-tium and cesium were driven off.

Not only could this account for the sharp rise in infant mortality in the northernmost part of the U.S. before the effects of the first bomb could make themselves felt, it could also explain the "Mandan Milk Mystery"—the inexplicably high strontium 90 content of the milk collected at Mandan, North Dakota, by the AEC's New York Health and Safety Laboratory throughout the 1950s. For the radioactive particles from the Hanford reactor in Washington would have largely passed over dry Idaho and Montana as they were blown by the prevailing westerly winds toward the wet eastern part of North Dakota where Mandan was located.

There was also one other peculiarity in the computer-printed maps. Florida, South Carolina, and Oklahoma showed no in-crease in infant mortality during the five years following the test, even though they had been in the path of the fallout. Soon, however, when I received detailed weather maps from the U.S. Meteorological Records Center, this too fitted the hy-pothesis: During the week ending July 17, 1945, the heavy rains had missed these states. In fact, the weater map indicated that the rainfall for that week had been restricted mainly to a rather narrow zone, extending from Texas along the northern edge of the Gulf of Mexico and then up the Atlantic coast. About 90 percent of the fallout is brought down by the rains, while only about 10 percent settles to the ground in dry air. This is why the zones of heaviest rainfall showed the sharpest increases. The *Weekly Weather and Crop Bulletin* summarized the rainfall situation as follows:

> Rainfall during the week was again limited mostly to sec-tions east of the Rockies. Heavy 24-hour amounts occurred the forepart of the week in Northeastern Texas and coastal areas of Gulf States and the latter part of the period in most Atlantic States and the lower Lake region.

And the summary ended: "Rain was light over the Central States and little or none occurred in the far west."

The *Weather Bulletin* also made it clear why the effects

of a single small bomb had been so serious. Referring to the rainfall, it added: "Haying continued in nearly all states with generally good to excellent yields."

Thus, the fallout from Trinity, which was twenty to thirty times greater than the fallout from later similar-sized tests because the fireball touched the ground and created enormous amounts of radioactive soil and vaporized rock, was deposited on the fresh vegetables and hay being harvested that week. And so the intensely radioactive short-lived isotopes, as well as the long-lived strontium 90, quickly found their way into the milk and food and from there into the unborn children in their mothers' wombs.

11

The Battle for Publication

EARLY IN JANUARY of 1969, the article on fetal and infant
mortality was returned by the editor of *Science*. This was the
paper that had been discredited by Abelson in his phone conversation with the reporter Stuart Brown, before the paper had
arrived in the offices of *Science*. Copies of three reviews were
enclosed. Two were clearly written by individuals in the field
of public health and statistics; these merely contained suggestions for certain changes that might make the case more complete. The third reviewer, however, was totally negative. This
individual went into a detailed analysis of fallout and dose
levels in the U.S., making eight references to internal AEC
reports. The majority of these reports had been prepared by
the staff of the AEC's Health and Safety Laboratory in New
York, directed by John Harley, the man responsible for the
classified fallout measurement at Troy.

The argument used to discredit the paper in this review
(and also in an article published by Harley in the *Quarterly
Bulletin* of the AEC's Health and Safety Laboratory later that
year) was this: According to the detailed measurements made
by the AEC's laboratory, the highest levels of strontium 90

had actually been in Utah, Montana, Wyoming, South Dakota, and Nebraska, as well as in very small portions of Massachusetts and Rhode Island. On the other hand, the lowest levels were measured in the southern U.S., from southern California to Florida. This was just the opposite of what I had said in my paper. Harley therefore argued that since infant mortality had risen most in the southern United States to the east of Nevada and New Mexico, and least in the low-rainfall mountain states of the Southwest, it clearly could not be the fallout that was responsible.

This was indeed a devastating argument, supported by a vast set of detailed measurements. Yet these measurements cited by Harley were in total disagreement with the measurements reported by the Public Health Service, which I had used in my paper. According to the Public Health Service, year after year the "wet" southern and eastern parts of the U.S. showed levels of strontium in the milk two, three, or even four times as high as in the dry western mountain states. The mystery was resolved a few months later when Dr. E. A. Martel, a former U.S. Air Force fallout specialist, told me the story of the gummed films.

The technique that had been used by Harley's lab to measure the fallout involved the use of a sheet of plastic about a foot square, coated on one side with a sticky substance very much like that used on flypaper. These plastic squares were mounted on a stand with the gummed side facing upward so as to catch the fallout particles as they descended. Every few days the films were collected, and shipped to the laboratory, where the radioactivity was measured.

In a detailed study later carried out by scientists at the Battelle Memorial Institute in Ohio, it was discovered that in the dry states of the west, the winds constantly picked up the radioactive dust again and again, so that the exposed gummed films, acting just like the flypaper in a room full of flies, ended up collecting much more radioactivity than was typical for the soil of the area. On the other hand, in the high-rainfall areas east of the Mississippi, the rains soaked the fallout deep into the soil and kept the dust levels low. Thus, by the early

1960s, it was widely realized that the so-called "gummed film" measurements of fallout had given levels much too high for the dry mountain states, and too low for the East and South. In fact, in the 1962 United Nations Report on Atomic Radiation, of which John Harley was one of the authors, there was a note from Harley on page 225 indicating that the "gummed film" procedure "may lead to an overestimate of the tropospheric fallout," the tropospehere being the lower part of the atmosphere containing the clouds of rain and radioactive dust.

Yet seven years later Harley used the "gummed film" measurements in his attempt to discredit my correlation of nuclear testing with the rises in infant mortality, writing that "fallout before 1954 was exactly the opposite of what was stated by Dr. Sternglass."

A few days after the paper on infant mortality had been returned by *Science,* the paper on the leukemia rise in the Troy area was also returned with a note of rejection. It was the same story again. Two of the three enclosed reviews were clearly by public-health physicians and statisticians and were quite favorable. One of these reviewers, in fact, stated that "the comparability between past fallout and past irradiated cases is 'impressive,' " while the other reviewer began with the statement: "The conclusion of this paper, if correct, is clearly a most important one." The third reviewer, however, was completely negative, and, exactly as in the case of the paper on infant mortality, there could be no doubt to which organization this third reviewer belonged. Almost word for word and point for point, the third review resembled a critique that had been sent to me a few months earlier by John Conway, Chief Counsel for the Joint Committee of Atomic Energy. And this critique had been sent to him by the AEC's Division of Biology and Medicine.

Perhaps the most remarkable point made by the third reviewer, in view of my experience with the Health Department of the State of New York, was the following: After arguing that the data on Albany-Troy were "incomplete," the reviewer asserted that "Sternglass could obtain the missing data."

And then, quite unexpectedly, less than a week after the

two papers had been returned by *Science,* the following letter arrived:

> *Bulletin of the Atomic Scientists*
> A Journal of Science and Public Affairs
> Eugene Rabinowitch, Editor

Dear Dr. Sternglass:

The drawings which we have for some of the figures in your article on Infant and Fetal Mortality Increase in the U.S. are not dark enough to be printed. Could you send us the original drawings—or very clear, dark copies—for figures 1 and 3.

Thank you for your assistance.

> Sincerely yours,
>
> Merry Selk
> Editorial Assistant

I gradually realized that the *Bulletin* must have decided to publish the report I had sent to Dr. Rabinowitch merely for his information.

As I later learned from the managing editor, Richard S. Lewis, in the face of strong reviewers' opinions both pro and con, it had been decided that the grave issues raised by my findings should be publicized and discussed as widely as possible, both by the scientific community and the general public. This was indeed good news. For although *Science* had a far wider circulation, the *Bulletin* reached an important group of physical and political scientists in university and government circles around the world. Furthermore, like *Science,* it was carefully read behind the Iron Curtain, hundreds of copies of each issue being sent to the Soviet Union, Poland, East Germany, Czechoslovakia, and Rumania.

Subsequently, I presented all the new evidence before a meeting of the National Council of the Federation of American Scientists, and the Council voted to set up a special committee to look into the evidence in detail. Dr. John T. Edsall, a noted biologist at Harvard University, agreed to head up the study committee, and after many months of investigations during which he consulted a number of specialists, he indicated in a letter that he would personally be willing to urge the editor of *Science* to reconsider his decision not to publish my findings.

Meanwhile, at the suggestion of a mutual acquaintance, I sent copies of my data to Dr. Luis Alvarez, then president of the Physical Society, who was now heading a physics research group in Berkeley, California. I included the maps showing increases in infant mortality downwind from the Trinity test site in New Mexico, for it had been Alvarez who watched the mushroom cloud from the first atomic explosion drift off across the United States. Thus, he was one of the few individuals in the world who had firsthand knowledge of the way the cloud had broken up and the directions in which it had drifted.

In the first paragraph of the reply I soon received from him, he stated that he had found the evidence "very impressive, particularly the map of the United States with the percent excess in mortality showing an effect only downwind of the Trinity site." He added that "in view of the enormous statistical significance of the results you plot on your map of the United States it is difficult to question your findings."

My article appeared in the April 1969 issue of the *Bulletin.* Interestingly, the managing editor, Richard Lewis, later told me that pressure had come both before and after publication in the form of long-distance calls from Washington from individuals who claimed to be long-term government friends of the journal. They said it was a grave mistake for the *Bulletin* to publish my article. When Lewis asked their names, they refused to identify themselves.

The April issue also carried an article by Dr. Freeman J. Dyson, a theoretical physicist at the Institute for Advanced Studies in Princeton, entitled "A Case for Missile Defense." It was clear that the basic premise of Dyson's argument in favor of an antiballistic missile (ABM) system would not be valid if my conclusions on the vulnerability of the developing infant to radiation were correct. His basic assumption was that a defensive system, once installed, regardless of how really effective it might be, would force an attacker to concentrate many of his missiles on a few defended cities, thereby reducing the number of cities that could be attacked with a given number of missiles and saving those cities that could not be attacked. But in the process, the "saved" cities would be inundated with intensive fallout.

I wrote a short note in rebuttal to Dyson's article, hoping that it would be published in the *Bulletin*. A few weeks later, a letter arrived from the *Bulletin* containing galley proofs of my letter and a reply by Dyson that began as follows:

> I welcome this chance to call attention to Ernest Sternglass' article "Infant Mortality and Nuclear Tests" in the April *Bulletin*. I urge everyone to read it. Compared with the issues Sternglass has raised, my arguments about missile defense are quite insignificant.
>
> Sternglass displays evidence that the effect of fallout in killing babies is about a hundred times greater than has been generally supposed. The evidence is not sufficient to prove Sternglass is right. The essential point is that Sternglass may be right. The margin of uncertainty in the effects of worldwide fallout is so large that we have no justification for dismissing Sternglass' numbers as fantastic.
>
> If Sternglass' numbers are right, as I believe they well may be, then he has a good argument against missile defense. . . .

Thus it appeared that once the evidence on the dangers of worldwide fallout was allowed to reach the scientific community at large, responsible scientists would be willing to reconsider their past judgments.

12

Counterattack
at Hanford

IN MAY 1969, for the first time in over fifteen years, as a result of growing concern among radiobiologists, there was to be a symposium dedicated to the effects of radiation on the developing mammal, including the human infant, both prior to and immediately after birth.

Not since the detonation of the first hydrogen bombs in 1953 had such a conference been sponsored by the Atomic Energy Commission. In the ensuing years, a vast amount of data had been accumulated on the biological effects of radiation given to animals at doses comparable to those that would be expected in a nuclear war. But even by 1969, almost nothing had been published on the more subtle effects of lower radiation doses, comparable to those from fallout, given over long periods of time. And there still had been no publication of any large-scale statistical studies of populations exposed to fallout.

It was true that the AEC had initiated a large-scale statistical study of some 20,000 persons exposed to long-term, low-level radiation in AEC laboratories. But the study population included only adults in their period of least sensitivity to radiation, receiving the best medical care available, and working under carefully controlled and monitored conditions designed to minimize any chance of absoring isotopes into their bodies. Since the control population consisted of individuals in the same families, communities, and installations who were not exposed to the radiation in the plant but who consumed the same fallout-contaminated food and inhaled the same fallout particles as the exposed test population, no effects from the fallout in the environment could in principle be detected by this study. More important, the study population did not include the children of the radiation laboratory workers. There was no search for unusual rises in stillbirths, infant deaths, congenital malformations, or cancer deaths among these children. When I inquired as to why this type of information was not sought, I was told that it was left out on the grounds that such questions might unduly alarm AEC employees.

There was clearly no need for the AEC to restrict its studies to so limited a population. In 1966 large statistical studies had been published regarding the very small amount of naturally occurring radium in drinking water in the states of Iowa and Illinois. In these studies a definite increase in bone cancer had been observed for the areas with high radium levels, clearly suggesting that the effects of very low levels of longterm radiation on large populations could in fact be detected, and, furthermore, that there was no evidence for a safe threshold even at the doses received from natural background radiation.

A letter from the Surgeon General of the United States, Dr. Jesse L. Steinfeld, to Representative William S. Moorhead of Pittsburgh made it clear that no large-scale epidemiological studies of possible low-level fallout effects were ever carried out or published by either the AEC or the U.S. Public Health Service, despite the fact that such studies had been specifically requested by congressmen Holifield and Price in the course of the 1963 congressional hearings and then promised by the Surgeon General then in office, Luther L. Terry. The exact words used by the present Surgeon General, Steinfeld, were:

. . . studies to determine the feasibility of a national program to analyze morbidity and mortality data of thyroid cancer, leukemia, and congenital malformations in relation to radiation exposure led to the decision that a national program was not indicated.

The letter added that "the feasibility studies were not published," and ended with the following statement:

> Some effects on human populations from levels of radiation of the magnitude encountered during nuclear warfare or from direct high-level radiation therapy have been well established. Effects from low-level radiation have not been as clearly delineated and further research into these problems is needed. The Service welcomes the opportunity to work with those who desire to construct and execute scientifically based plans for studies of human effects from low-level radiation exposure.

Thus, in all the years since 1963 no large-scale study of the effects of low-level radiation from *fallout* had evidently been undertaken.

The Surgeon General's letter did say that studies of fallout-exposed schoolchildren in Utah were being conducted by Dr. G. D. Carlyle Thompson of the Utah State Department of Health and that Dr. Thompson had published some preliminary data on thyroid cancers among these children in the October 1967 *American Journal of Public Health.* "Progress reports to date," wrote Surgeon General Steinfeld, "show no unusual increased incidence of leukemia and no cases of thyroid cancer among children who reside in the selected 'exposed' area of Utah-Nevada."

Examination of Thompson's article, however, showed that, although Steinfeld's statement as phrased was literally true, it was misleading in its implication. For although Thompson's paper indeed indicated no increase in thyroid cancers in these children relative to their counterparts in New York State when both sexes were combined, there was in fact a significant increase among girls zero to 19 years of age. Among this group there was a total of ten cases for 1958–62, as compared to four cases expected. For young women aged 20 to 29 years old, the number was twenty as compared to nine expected.

Still more significant, the rate of thyroid cancers per 100,000 young women under age 30 in Utah had increased almost 400 percent—from 0.6 in 1948–52, before the Nevada tests, to 2.3 in 1958–62.

Since no figures on leukemia deaths in Utah were given in Thompson's article, it was necessary to consult the U.S. Vital Statistics, and what they showed was quite unbelievable. For the age group 5 to 14 years, there were large percentage rises in leukemia deaths exactly three to five years after each of the major test series that deposited fallout in the Utah area. Between 1949 and 1967, the annual number had increased fourfold in successive peaks from 1.5 to 6.2 per 100,000 children. But since leukemia rates for children 5 to 14 years in New York State and elsewhere also went up when fallout became widespread, although not as much as in Utah, a statistician could perhaps say that there was "no unusual increase of leukemia" in Utah.

This, then, was the situation in May 1969 on the eve of the Hanford conference. According to the preliminary program of the symposium, Dr. Alice Stewart would be present. The conference was to take place at Hanford, Washington, the site where the plutonium for the Trinity bomb had been produced in 1944. A few days before I was scheduled to leave for Hanford, I watched the computer print out map after map of rising infant mortality stretching eastward from the Hanford reactors and from the Trinity site at Alamogordo, New Mexico.

At about this time, a completely unexpected letter arrived by special delivery from the New York State Health Department's Bureau of Cancer Control:

April 30, 1969

Dear Dr. Sternglass:

Doctor James Yamazaki of the University of California School of Medicine has written to me about your approaching presentation at the Ninth Annual Hanford Biology Symposium. In his letter, Doctor Yamazaki requested information about environmental factors in the Albany-Schenectady-Rensselaer country areas in the early 1950s.

As you can see from my response, a copy of which is enclosed, this Bureau has reviewed some of the data upon
which your June 1968 presentation in Denver was based.
The results of this preliminary analysis are noted in the
letter to Doctor Yamazaki. It is my opinion these data
are too inaccurate for the type of analysis you have done.
We are planning to restudy this subject and would be happy
to have you take part.

I hope this reaches you before the May 5th conference.

> Sincerely yours,
> Peter Greenwald, M.D.
> Director
> Bureau of Cancer Control

The enclosed letter which Greenwald had written to Yamazaki indicated that J. H. Lade of the New York State Health
Department had apparently made some errors in his data on
leukemia in Albany-Troy, published years before in *Science*
in the attempt to prove that there had been no effect from
the fallout of April 1953. As Greenwald put it, "While it is
unfortunate that this Health Department may have erred in
not clearly describing the possible inaccuracies in this table it
is clear that they should now be pointed out." Upon examining
Greenwald's revised table, I discovered a remarkable coincidence: These newly found inaccuracies concerned only the cases
born in 1953, the year the fallout arrived. Greenwald's new
figures showed a greatly reduced number of leukemia cases
in that year, thus making it now possible to argue that there
had been no significant effect from the fallout.

At the end of Greenwald's letter was the notation:

cc: Dr. Sternglass
    Dr. Sagan

And on the preliminary program of the Hanford meeting, listed
below my name, was:

9:40 P.M. INVITED DISCUSSANT
           L. A. Sagan
           Palo Alto Medical Clinic

This was Dr. Leonard Sagan, who had long been working for the AEC's Division of Biology and Medicine. He was apparently scheduled to be discussant, or critic, for my paper at Hanford. Why was he listed as being affiliated only with the Palo Alto Medical Clinic? Interestingly, my paper was the only one in that session for which a discussant had been listed in the final program. The co-chairman of the session was Dr. J. N. Yamazaki of the University of California, the same individual who had just requested information on the Troy rainout from the New York State Health Department. Dr. Yamazaki had published a number of papers while working for the Atomic Bomb Casualty Commission that argued for the absence of significant leukemia rises among offspring born to the survivors of Hiroshima.

Upon arriving at Hanford, I was interested to discover the summary of a paper written by Dr. Y. I. Moskalev and his colleagues at the Institute for Biophysics in Moscow. This group of Russian scientists had studied the effects of low-level radiation—of the type produced by fallout—from strontium 90 and other isotopes when given to various animals before and during pregnancy. Their observations on the offspring of all types of animals, ranging in size from rats, rabbits, and dogs to sheep and cows, appeared to show effects similar to those I had postulated for human infants. There was no spectacular increase in gross malformations at low levels of strontium 90. Instead there was a small reduction in weight at birth, a decline in fertility, and an increase in the number of fetal deaths. And the earlier the radioactivity was fed to the pregnant animals, the more pronounced were the effects, just exactly as the studies of diagnostic X-rays given during pregnancy had indicated in the case of humans.

Unexpected difficulties, however, developed, and the Russian scientists were unable to attend. This was a great disappointment, for the studies of Dr. Moskalev and his group were the first published work I was aware of that appeared to offer the kind of crucial laboratory confirmation of the statistical evidence for the damaging effects of fallout on the newborn. Early in the conference, however, other evidence was presented, in session after session, that internal radiation from isotopes of the type that occurred in fallout was particularly hazardous to the ova and the early embryo and fetus. These were levels much lower than those that led to lethal effects in mature animals.

First on the program on the evening my paper was scheduled was a paper by Dr. M. L. Griem of the University of Chicago School of Medicine. Dr. Griem indicated that in his careful study of children X-rayed while in the womb there was no evidence for a difference in the number of leukemia cases between the irradiated children and the two control groups who had not received any radiation. But as soon as Dr. Griem had begun his talk, it became apparent that the number of children X-rayed was only 1008, a group in which one would normally not expect more than one case of leukemia in ten years of life. Thus, even a doubling would at most result in one extra leukemia case, clearly too small an effect to be readily observed with any statistical certainty. As Dr. Alice Stewart herself pointed out in the ensuing discussion, "the reality of juvenile leukemia is such that no one could hope to do a follow-up like this and detect an increase of 50 percent of the normal incidence without being prepared to follow out 900,000 children for ten years, and no one has done that." She herself had to interview the families of over 7000 children who had died of cancer out of a total of nearly nineteen million children born in order to establish a clear causal connection.

Interestingly, however, Dr. Griem's study actually lent support to the hypothesis of other effects from low-dose radiation, even in so small a study population as 1008 children. His data showed that, although there was no detectable increase in leukemia, there was a significant increase in benign tumors and certain types of congenital birth defects, especially severely disfiguring birthmarks. Furthermore, Griem's study revealed that stillbirths had nearly doubled—from eight and nine cases in the two control groups to sixteen among those who had received X-rays in the womb. And according to Griem, the X-rays were given mainly in the less sensitive second and third trimesters of pregnancy, with doses of only 1 to 3 rads to the fetuses. But Dr. Stewart had found the sensitivity to be some ten times greater in the first three months of pregnancy. Thus, Griem's study indicated that if the X-rays had instead been given in the earliest developmental phase, a dose only about one-tenth as large, or a mere 100 to 300 millirads, would have produced a doubling of stillbirths and genetic defects. These were, in fact, the general doses that had been received by large numbers of unborn children from fallout. So Griem's study actually provided, for the first time, direct observational evidence in humans for my statistical findings on fetal mortality.

Soon it was Dr. Stewart's turn. As her opening paragraph made clear, it was the opposition to her findings that drove her to find more and more convincing evidence:

> More than ten years ago, a retrospective approach to the problem of cancer etiology led to the conclusion that about one in twenty of the 600 cancer deaths that were taking yearly tolls of the seven million children living in Britain were being caused by obstetric X-ray examinations. So far as the investigators themselves were concerned, the matter would probably have rested there had it not been for the general skepticism that greeted this suggestion.
>
> However, so many clinicians and experimentalists refused to accept the necessarily epidemiological evidence that it was finally decided to remove all grounds for doubt by isolating groups of mainly radiogenic cases (i.e. X-rayed cases) and wholly non-radiogenic cases (i.e. non X-rayed cases) and observing the ages of the two groups of children when symptoms developed.

She then went on to present new evidence that leukemia cases among children X-rayed while in the womb showed a clearly recognizable, narrow age range of between three and five years after birth, evidence that supported the hypothesis that the excess cases in Albany-Troy among the children irradiated in the womb were caused by radiation, since they too showed the same abnormal pattern of age at death.

Dr. Stewart also observed that all types of cellular defects produced by radiation "were felt with equal force by all systems and organs" in the fetal stage, a point crucial in explaining how radiation could cause subtle types of damage that would lead to such biochemical defects as the inability to resist infections early in infancy or childhood. And in her conclusion, she pointed out that her evidence clearly showed that the effects that prove fatal before the age of 10 years are mainly initiated before birth, and often in the most sensitive period of early development. Thus Dr. Stewart's paper had greatly strengthened my statistical findings on the effect of fallout radiation.

I then presented my evidence, concluding with the estimate that some 400,000 infants of less than one year of age probably had died as the result of nuclear fallout between 1950 and 1965. The chairman then announced that there would be no discussion until Dr. Sagan of the Palo Alto Medical Clinic had commented on the paper. Dr. Sagan took the microphone.

He began by showing the slides of the material collected for him by the co-chairman of the session, Dr. Yamazaki, from the New York State Health Department. Sagan made the point that the data were too unreliable for the type of study I had done. He next cited an estimated figure for the expected number of leukemia cases in a population exposed to a radiation dose of 1 rad, namely one case per year among each million persons. Therefore, he concluded, the doses in the Troy area, with its 130,000 children, were some 1000 times too small to produce even a single extra leukemia case. Sagan's estimated figure, however, was the one for the mature adult, who is least sensitive of all to radiation. It was evident from Dr. Stewart's and Dr. MacMahon's work, as well as from the paper just delivered by Griem, that the fetus and especially the early embryo were many hundreds of times as sensitive as the adult. Furthermore, the total dose from internal radiation in Troy was at least five times the external dose measured by the AEC's Health and Safety Laboratory—which Sagan had used in his argument.

He next cited Harley's invalidated "gummed film" fallout measurements to discredit any association of fallout with the infant mortality rises in the wet areas of the South and East. He did this despite the fact that the Public Health Service's data for strontium in the milk that I had just presented correlated perfectly with the changes in infant mortality in every state for which the Service had published data.

The discussion was then opened to the floor, and a long

and often heated debate began. Some AEC scientists had slides that showed that miniature swine fed fairly large doses of strontium did not seem to show detectable reductions in weight and litter size. This was in apparent contradiction to the findings of Moskalev, who was unfortunately not pesent to defend his work. But then Dr. T. K. Ellis of the University of Utah reported that in beagle dogs given small amounts of strontium 90, surprisingly strong effects on male hormone production had been found, as well as changes in reproductive cells. Dr. Harold Rosenthal, who had made measurements on baby teeth collected from all over the United States, indicated that for Texas, St. Louis, Toronto, Detroit, Chicago, and California he had found a close correlation between levels of strontium in the teeth and the levels reported for the milk. This showed that the Public Health Service's milk measurements that I used in my work were indeed a good indication of the amount of exposure received by infants in various areas.

The evidence supporting unexpectedly severe effects on the early embryo and fetus from internal radiation sources continued to accumulate until the close of the conference. And afterward, when I was able to examine the original manuscript submitted by Dr. Moskalev, I found that his studies did in fact confirm the most crucial points at issue in the whole fallout controversy. His group had found that when various of the isotopes contained in fallout were fed to female animals during pregnancy, large fractions were transferred through the placenta to the developing fetus. For example, up to 38 percent was transferred in the case of strontium given to rats and up to 66 percent in the case of cesium given to a litter of dogs. Furthermore, the amounts reaching the developing fetus were many times greater for continuous, slow intake (such as occurs with fallout in food) than for a large single dose.

More significant, Moskalev had found a direct relationship between the size of the doses of isotopes given just before pregnancy and the percent of the offspring that died—even for doses as small as 4-billionths of a curie per gram of body weight. This dose was well within the range of doses from fallout delivered to the early human fetus by the accumulated strontium in the mothers' bones. Thus, the argument that there might not be any effects at all from long-term, low-level radiation as opposed to doses given all at once, like diagnostic X-rays, had now been disproven by direct experiments.

Moskalev's results also showed that, regardless of whether a given amount of isotopes was fed to a mouse weighing 20 grams or to a dog weighing many thousands of grams, approximately the same fraction of the total amount would always concentrate in the rapidly growing embryo. This was one reason why, in all species, the fetus was so much more sensitive to fallout radiation than the adult. A given tiny amount of an isotope in the body of an adult might be quite tolerable if it was evenly distributed throughout the 70,000 grams of an average woman's body weight. But if even only one one-hundredth of this amount in the mother's body goes to the embryo during the first two to four weeks of development, when the embryo weighs less than a hundredth of a gram, then the concentration of radioactivity in the embryonic tissue would be the same as if the entire original amount had been given to a 1-gram embryo. Instead of the relatively minor radiation dose produced by this original amount distributed in the 70,000 grams of body weight typical of the adult, there would be a concentration 70,000 times greater in the early embryo. And added to this was the fact that the embryo was already hundreds of times more likely to develop cancer or other forms of biochemical damage than the much more resistant, fully mature adult, as paper after paper presented at the conference showed. Thus, the low external doses given by fallout to the body of an adult, doses on which the world's radiation protection agencies had based their assumption that fallout was harmless, were actually highly lethal doses for the early embryo.

It was not our physics and technology that had been inadequate, but our knowledge of biological systems and their enormous ability to concentrate toxic agents. Just as in the case of DDT, it was not the amount distributed throughout the environment that was so serious. It was the selective concentration in the food chain and then in the newly forming organs of the rapidly developing young embryo. Since all higher animals, including man, must pass through this critically sensitive phase, it was clear that, unless the problem was widely recognized and acted upon, man could extinguish himself and all other animals, not through the effect of radiation on the adult, but through the effect on the weakest link in the chain of life— the unborn and the very young.

13

The Public's Right
to Know

NOT A WORD of the findings reported at Hanford appeared in the national press or the other public news media. No science writers had been present, and none of the wire services had covered the meeting, held in a remote part of the country. Yet in view of the accumulated evidence presented at Hanford, it was now clear that the sensitivity of the early embryo was so great that if a nuclear war ever broke out only the more resistant reptiles and insects would survive the lingering radiation.

In June 1969 I wrote a letter to *The New York Times* describing the Hanford findings and their implications for the ABM debate. Shortly after the letter appeared, I received a phone call from the New York correspondent of *The London Observer*, Joyce Egginton. The first thing Miss Egginton asked was why there had not been any stories about these startling findings either in the news section of *The New York Times* or in any other newspapers served by the national wire services. In particular, she wondered why nothing had been reported about my findings in view of the fact that I had just presented them again at a recent meeting of the Health Physics Society. There, the University of Pittsburgh's public relations department had made arrangements with the public relations department of the Health Physics Society for an interview with an individual who stated that he wanted to write a story on my work for the Associated Press. After the presentation of my paper, this individual introduced himself, stating that he was in charge of the newsroom for the Society but also often wrote stories for the Associated Press. He already had much of the story written and merely wanted to check certain details. It seemed to be quite a good account. But his story was never carried by the Associated Press. Not until later did it become clear what had happened.

Meanwhile, the editor of *Esquire,* Harold Hayes, called soon after Joyce Egginton to ask whether I would be willing to write a story on my findings for his magazine. In view of the fact that the general public had heard nothing at all of the findings presented in Pittsburgh or at Hanford, I accepted.

A week or two later, I received a call from Hal Stromholt, a writer for the Associated Press office in Pittsburgh. He said that the AP had asked him to do a story on my findings presented at the recent Health Physics Society meeting, which Joyce Egginton had just reported in *The London Observer.* I asked him why the AP had not carried the story that had been written for it many weeks before. Surprised, Stromholt asked the name of the man who had done this earlier story. I told him, and he then said that no one by that name either worked in the AP office in Pittsburgh or was employed as an occasional writer or "stringer." Furthermore, he added that no earlier story could have been sent to the AP, for they certainly would not have asked him to write a second one on exactly the same news item. Significantly, the Associated Press did in fact use Stromholt's story, and it was carried throughout the country.

Shortly thereafter the publishers of *Esquire* decided to stop the press run on the September issue in order to include the article on infant mortality and nuclear testing as a special insert. Furthermore, in view of the serious implications for the decision on whether the U.S. should build the new antiballistic missile nuclear defense system, which would fill the atmosphere with hundreds of times as much fallout as all of the past nuclear tests if it were ever to be used, the publishers had decided to take out full-page advertisements in *The New York Times* and the *Washington Post* that would summarize the principal points of the article. They felt that these advertisements might appear in time to be considered in the Senate debate. Harold Hayes also sent advance copies of the article to every congressman and senator, together with a personal letter explaining the reason for this unprecedented action on the part of his magazine.

But for the ABM debate, it was too late. The final vote came only eight days later with a narrow victory for the Defense Department and the AEC, before the evidence of biological risks had a chance to be fully considered by Congress and the public.

On October 12, 1969, a few weeks after the *Esquire* article appeared, *The New York Times* published an account of a critique of my Troy data prepared by Dr. Peter Greenwald, director of the New York State Health Department's Bureau of Cancer Control, and Mrs. Sandra Kinch, director of the department's Office of Biostatistics. According to the *Times,* these two officials said that their analyses "tend to refute the validity and the conclusions" of my studies, which they described as having "definite factual errors" in the data used. To prove this, they submitted their own table of leukemia deaths among children under age 10 in the Albany-Troy-Schenectady area. The figures they gave were as follows:

Birth Years	No. of Deaths
1940–42	6
1943–45	8
1946–48	9
1949–51	7
1952–54	10
1955–57	13
Total:	53

But the figures supplied by Dr. J. H. Lade in his 1964 letter to *Science,* which I had used in my study, were (when regrouped in similar three-year periods):

Birth Years	No. of Deaths
1943–45	9
1946–48	8
1949–51	9
1952–54	15
1955–57	13
Total:	54

This was indeed peculiar. Five cases were missing from Greenwald's table in the critical 1952–54 period, yet the overall total number of cases was nearly the same in both tables. Examination of Greenwald's table showed that he had added a new category not present in Lade's: the 1940–42 period, for which he listed six cases. He had also deleted two cases from the 1949–51 period, with the final result that the total number of cases was nearly the same for both tables—even though the exact five cases that indicated a large increase in leukemia during the critical years 1952–54 were missing.

Why had these five cases been removed? I re-examined the correspondence I had received from Greenwald. In it he stated that inaccuracies had been found in Lade's table among five of the eight cases born in the year 1953. But should these five cases have been eliminated from the 1952–54 entry? The answer could be found by examining Greenwald's critique, in which he described the nature of the inaccuracies in these cases. He stated that two of the children had actually been born in 1952 instead of 1953, while another had been born in 1954. But in his table in the *Times,* Greenwald had used the time period 1952–54, *not* 1953. Therefore these cases should have been included in his table by his own criteria. The fourth missing case was born in Montgomery County, New York, according to Greenwald, and not in the Albany-Troy area. But Montgomery County is right next to Albany-Troy, so this case could certainly have been caused by the fallout. And the same would hold true for the fifth and last child, who was born in New Mexico, according to Greenwald, and brought to Troy as an infant, and thus also would have been exposed to the fallout in the diet. I had hypothesized in my study that the fallout could have caused leukemia by genetic damage to the parents before conception or by direct damage either to children in the womb or to young infants. Therefore, there was no reason to exclude *any* of the five cases from the evidence. If they had been included, Greenwald's table would have shown fifteen cases for 1952–54, a doubling over the average rate for the previous years. Furthermore, the thirteen cases in the 1955–57 period were consistent with the hypothesis of genetically caused leukemia.

Other distortions of data were present in the critique by Dr. Greenwald and Mrs. Kinch. But how would the general public ever suspect? This was the voice of the New York State Health Department and not the AEC. Readers of the nation's press would simply assume that responsible and independent public health officials had proven that fallout was harmless after all.

During this period, amid the resultant publicity surrounding the article in the *Observer,* I was invited to appear on the NBC-TV "Today" show. According to Hugh Downs, one of the reporters on the program, the AEC had learned of the plans for my appearance and called the producer, urging him not to invite me. When the producer refused, the AEC urged that a scientist holding an opposing view should be present to give an immediate rebuttal. When this also was turned down, the AEC insisted on equal time as soon as possible after my appearance so that two independent scientists would be able to present the argument against my thesis. NBC finally agreed to provide equal time the following week.

On the program itself, Hugh Downs brought this out into the open: "I was just going to say that the Atomic Energy Commission called us yesterday. They were concerned about your appearance on the program today."

I asked the identity of the independent scientists whom the AEC wanted to present. Downs replied that one was a physician by the name of Dr. Leonard Sagan from the Palo Alto Medical Clinic, and the other was a Dr. John Storer from the Oak Ridge National Laboratory. Both, of course, had worked until very recently for the Atomic Energy Commission's

Division of Biology and Medicine. As to the reason why the
AEC had urged NBC to cancel my appearance, Hugh Downs
reported: "They say that their experiments with animals show
that there is no damage to offspring at all from parent animals
given strontium 90 in those low dosages that we get." This
was in complete contradiction with the evidence submitted at
Hanford by Moskalev and numerous other independent scientists. The explanation lay, as usual, in the manner in which
the AEC studies were constructed.

This point was to be confirmed a few months later in a
most unexpected manner by a chief scientist at one of the AEC's
own laboratories. I had received an invitation to present a paper
on my other research work regarding the reduction of diagnostic X-ray doses at a meeting in San Francisco in October 1969.
I was also asked whether I would be willing to debate my
fallout thesis on the Berkeley campus with Dr. Arthur Tamplin,
who had written a critique of my work that was about to be
published in the *Bulletin of the Atomic Scientists,* and to which
I had just written a reply. Tamplin was a biophysicist at the
nearby Livermore Laboratory, operated for the AEC by the
University of California. During the question period following
our debate, someone in the audience brought up the argument
that AEC studies had found no significant increase in mortality
among the offspring of experimental animals fed strontium 90
for long periods of time. Immediately, someone else in the
audience stood up and asked to comment on this question. It
turned out to be Dr. John Gofman, Tamplin's supervisor, who
was Director of the Biomedical Division of Livermore and
an Associate Director of the laboratory. This was the man
the AEC had placed in charge of all their radiobiological studies
at Livermore back in 1963, when the hazard from internal
fallout doses first aroused widespread public concern. For years
Gofman had been studying the possible connection between
radiation, chromosome defects, and cancer. He told the audience that he had investigated all the animal experiments carried
out by the AEC, and in no case had they been designed to
detect the kind of small reduction in birthweight and ability
to fight infections that I had suggested as the likely cause for

the increased infant mortality in man. He concluded that, to the best of his knowledge, there was not a single animal experiment that would contradict my hypothesis, and with that he sat down.

Within less than a year, both Gofman and Tamplin publicly denounced as complete falsehood the position of the AEC as expressed by Sagan and Storer on the "Today" program, namely, that "the levels of radiation to which the American public was exposed from fallout have been harmless." As told by the two scientists themselves in the pages of *The Bulletin of the Atomic Scientists,* their public denunciation of the AEC's position on low-level radiation effects was precipitated by the attempt of the AEC's top management to force Tamplin to suppress his own independent calculations, made in his original critique of my findings, that perhaps as many as 8000 infant deaths per year might have taken place as a result of genetic damage from nuclear testing. Dr. Spofford English, the Assistant General Manager in charge of the AEC's entire research program, together with the head of the Division of Biology and Medicine, Dr. John Totter, as well as Dr. Leonard Sagan and Dr. John B. Storer, had indicated to Gofman that Tamplin should publish his critique minus his own estimate of the possible fetal and infant deaths, thus effectively keeping this information from the public.

As Gofman and Tamplin put it: "They wanted us by *omission* to support their incredible position as stated on the 'Today' show, and to put Tamplin's estimate into a less widely read scholarly journal, where it would evidently not be seen · by the scientific community at large, the general public, and their elected representatives in Congress."

Subsequently, Dr. Gofman resigned his position as Associate Director of the Livermore Laboratories, and all but one of Tamplin's research group of twelve people were taken away from him six months later. Both scientists have continued to testify before various congressional committees that there is no safe threshold of radiation exposure and that presently permitted radiation exposure levels must be cut back to virtually zero. They have proposed that no release whatsoever of radioactive materials into the environment should be permitted without a full, nonpartial, interdisciplinary examination of each situation. And they have assembled a vast body of data indicating that if the radiation doses now allowed by AEC regulations (an average of 170 millirads per year for the entire population and not more than 500 for any single individual) were to be received by the entire U.S. population as a result of peacetime uses of nuclear energy, there would be at least 32,000 and perhaps as many as 64,000 additional deaths each year from cancer and leukemia alone. And these figures did not even include fetal and infant mortality or any more subtle long-range effects on health.

At the hearings on the environmental effects of electric power generation held by the Joint Committee on Atomic Energy in November 1969 and January 1970, Gofman and Tamplin presented their conclusion that a direct relationship exists between low-level radiation doses and the development of cancer, not only in the fetus and infant, but also in the mature adult. Furthermore, they urged an immediate tenfold reduction in the permissible radiation doses to the general population from all peaceful uses of nuclear energy.

14

The Price of Secrecy

As LATE AS the spring of 1970, I believed that the radiation resulting from the normal operation of nuclear power plants was so low as to present no significant hazard to public health. This belief was based on the results of an old study of emissions from the first commercial nuclear electric power plant, located at Shippingport, Pennsylvania. Conducted almost ten years earlier, it had measured the radioactivity in the cooling water taken from the Ohio River both before and after it had passed through the plant. It found that the plant added so little radioactivity to the water that there were times when the chemically filtered water leaving the plant was actually less radioactive than the river water entering the plant—especially during periods of heavy fallout.

It therefore seemed reasonable to expect that if such low levels of radioactive waste releases had been achieved in 1957 in the very first nuclear power reactor built in this country, then the later, more advanced plants would release even less. But early in 1970 I discovered that this was not the case. In the published record of the hearings on the environmental effects of electric power generation, held by the Joint Committee on Atomic Energy in November 1969, there were tables supplied by the AEC listing the amounts of radioactivity discharged into the water and air by commercial nuclear power plants in the United States. Many plants were listed as actually releasing hundreds of thousands of times as much radioactivity into the air as others. For example, in 1967 two reactors had discharged as much as 700,000 curies, while another had released only 2.4 curies, or some 300,000 times less.

These were truly enormous quantities. Some of the many different isotopes contained in these gaseous and liquid discharges, such as cesium and strontium, were regarded as hazardous at levels as low as one ten-billionth of a curie per day in milk or food. A single curie of iodine 131 could make 10 billion quarts of milk unfit for continuous consumption, according even to the existing guidelines adopted by the federal government. Such large releases of radioactivity were in fact comparable to fallout from small tactical nuclear weapons. Although dilution in the air would reduce the hazard to people living more than fifty miles away from these plants, those living nearby were unknowingly accepting vastly greater risks to the health of their children.

Furthermore, the *permissible* levels listed for many of the reactors were enormous. For the Dresden reactor, located some fifty miles from Chicago, which had emitted 260,000 curies of radioactive gases in 1967, the permissible amount had been set at 22,000,000 curies per year by the AEC. Thus, in terms of permissible levels, the huge amount actually released could be, and was, cited by the power company as representing only about 1 percent of the maximum levels allowed.

Curiously, there was no listing given in the record of the hearings for the Shippingport plant. Why had the AEC left it out of the material presented to Congress? A few months later, at a meeting of the Health Physics Society, Charles Weaver, Director of the Division of Environmental Radiation of the Public Health Service's Bureau of Radiological Health, presented the results of studies just published in March of 1970 on the radioactivity emitted by a series of nuclear plants. According to his report, in 1968 Shippingport had emitted a grand total of only 0.001 curies into the air, or 240 million times less than was released the same year by the Dresden reactor near Chicago, Illinois. And the same report also explained why some reactors released so much more radioactivity than others. They were grouped according to type of design, and it was only the boiling-water-type reactors that showed such very large releases, while the pressurized-water reactors, which included Shippingport, consistently showed the lowest waste discharges. Some of the commercial pressurized reactors did, however, emit much more than others, and all of them discharged significant quantities of radioactive tritium into the cooling water, which was then released into the surrounding rivers and lakes.

Why weren't all the reactors designed like Shippingport, so as to release the smallest amounts of radioactivity? The answer could be found in the history of reactor development. The pressurized-water reactors, like Shippingport, were originally designed for use in nuclear submarines by Westinghouse under the direction of Admiral Hyman C. Rickover. Since they had to operate for long periods in a sealed, submerged vessel, these rectors had to be designed with a minimum of radioactive leakage either into the submarine, where the crew had to live for months as a time, or into the water, where the bubbles of radioactive gases would permit easy detection of the submarine's position. The Shippingport reactor was in fact a prototype naval propulsion plant owned by the Navy and the AEC, and not a commercial power plant at all.

Meanwhile, the General Electric Company was encouraged by the AEC to quickly develop a new type of large power reactor that would be cheap and efficient enough to compete successfully with the fossil-fuel-burning electric power plants in widespread use. For the more complex pressurized reactor with its double cooling loop, although safer, was too expensive. And so GE developed the much simpler boiling-water reactor. This design, in which economic considerations were the major factor, sacrificed protection against radioactive leaks in favor of lower cost and greater efficiency of operation. Experiments showed that corrosion was a more serious problem in the single-coolant-loop GE design. Large amounts of fission products would inevitably build up rapidly in the coolant and escape through pipe joints, valve packings, and high-speed rotating shaft-seals to be discharged into the air and water. Thus, if a cheap, economical way to generate large quantities of electric power was to be demonstrated quickly so as to convince the utilities to go nuclear, there was only one solution: Set the permissible amounts of radioactive waste discharges into the environment so high that the actual releases would always be well below this limit.

By 1959, the first large boiling-water reactor plant was completed at Dresden, Illinois, and in August of 1960, the first electricity from the 200-megawatt Dresden generators began to flow into the power grid of the Commonwealth Edison Company, serving the people of Chicago. The releases of radioactive gases into the atmosphere were relatively low in the first full year of operation, and so were the discharges of tritium, strontium 90, and other isotopes into the Illinois River. In 1961, only 0.158 percent of the maximum allowable amount had to be released into the air, and even liquid wastes were held down to 6.3 percent of permitted levels. Compared to the amounts of radioactivity then being released into the water and the air by the renewed testing of nuclear weapons, this was certainly quite small.

But signs of trouble began to appear the very next year. By the end of 1962, corrosion had begun, and the amount of radioactive gas that had to be discharged into the air increased by almost ten times to 284,000 curies. Even the radioactivity discharged into the river rose more than three times. By 1963 emission of radioactive gases had been successfully brought back down to 71,600 curies by the replacement of leaking fuel rods, but the corrosion continued, and gaseous releases shot up to 521,000 curies in 1964.

No longer were the radiation doses to the surrounding population negligibly small compared to background radiation, as everyone had hoped. Annual average external doses to the population within a few miles of the plant could be estimated at 20 to 30 millirads by 1964. This was fast approaching the 88 millirads that the people in the area normally received from cosmic radiation and natural radioactivity in the soil, and it compared with what had been produced by weapons fallout.

It was becoming clear that the permissible levels of radiation from nuclear plants could not be lowered, as some scientists were beginning to urge, without having to shut the huge plant down only a few years after it had been built at a cost of well over a hundred million dollars. In fact, pressures were actually building up from industry and the military to *raise* the permissible discharges to the environment from nuclear activities, especially in the event of an accidental heavy release from a reactor or from fallout if weapons tests in the atmosphere were ever renewed. And so, in 1964 and 1965, the director of the Federal Radiation Council, Dr. Paul C. Tompkins, who had previously served as Deputy Director of the AEC's Office of Radiation Standards and Director of Research in the Bureau of Radiological Health of the U.S. Public Health Service, announced a twentyfold rise in the permissible amounts of the most hazardous isotopes in milk in the event of an accidental release. For the first time in the history of radiation standards the permissible doses to the public were raised rather than lowered, despite the mounting evidence that there was no safe threshold dose of radiation as presented in August 1963 before the Joint Committee. And this was done quietly by presidential executive order, for which no public hearing is required.

When in 1966 the gaseous discharges from the Dresden plant had climbed to 736,000 curies, or more than twenty times what they had been in 1961 and more than twenty million times more than Shippingport had released the same year, a decision was made to start replacing the corroding stainless-steel-jacketed fuel rods with more resistant, but also more expensive, zircalloy-clad fuel. By this time, the liquid releases, containing iodine, strontium, cesium, and other highly toxic elements, had risen to forty-three times their initial value, and, instead of being a small fraction of the permissible level, they had actually reached a full third of the AEC standards. Enormous quantities of these isotopes went into the Illinois River, flowing past Peoria, where the river water began to be used for drinking, and on to the Gulf of Mexico, concentrating thousands of times higher in the fish and in the birds that fed on them.

The example of Dresden clearly showed that it would not be possible to lower permissible radiation levels without having to shut down the whole series of boiling-water reactors that had now gone into operation all over the United States, each having cost some one hundred million dollars. And construction would have to be halted on dozens of even larger reactors in various stages of development throughout the United States and the rest of the world.

On June 6, 1970, just a few months after all the new reactor emissions data had been published, the British medical journal *Lancet* printed a full account of Dr. Stewart's fifteen-year study of the increase in leukemia and cancer among the nineteen million children in England and Wales that were born between 1943 and 1965. Her conclusions were now statistically unassailable: doubling the number of X-ray pictures doubled the risk of leukemia and other cancers, and there was no evidence for a safe threshold even at a single diagnostic X-ray. One modern pelvic X-ray gives about the same dose to the fetus as is permitted for the general population by the existing federal radiation standards (175 millirads). And when the radiation was given in the first three months of pregnancy, Dr. Stewart's data showed that a mere 80 millirads—about the dose that was received from external radiation alone by the people living near the Dresden reactor in the peak year of emission—would double the spontaneous rate of leukemia and cancer in the children before they reached the age of ten. And there had been significant rises in leukemia and fetal and infant mortality in the Troy area at similar external dose levels of only 50 to 100 millirads.

There was thus little doubt that detectable health effects should have occurred in the areas surrounding the Dresden plant and other reactors. But it would require a considerable effort to collect the data from the volumes of the U.S. Vital Statistics, and I had no one to help in such a task, for I had been unable to obtain any funds for such studies. Fortunately, a group of students who had become interested in environmental pollution as a result of Earth Day indicated a willingness to help. The group divided itself into small teams, and each took on the task of gathering the data for a particular nuclear installation.

Since most of the fission products emitted by reactors are short-lived, persisting only for anywhere from a few days to a few months, it appeared that the effects on infant mortality would be sharp and immediate, just as had occurred with the short-lived isotopes in the case of fallout. There would probably be no significant residual effect, and so the rises and falls in infant mortality should correlate closely with the rises and falls in the reactor releases.

In October 1970 we examined the infant mortality rates in the counties around the Dresden reactor. In 1966, within a year after the emissions rose sharply from the relatively low value of 71,600 curies in 1963 to 610,000 curies in 1965, the infant mortality rate in Grundy County, where the reactor was located, and in adjacent Livingston County, jumped by 140 percent, or to more than twice its 1964 value. While only thirteen infants in these two counties had died in the year after the minimum radioactive emission, by 1966 this number had jumped to thirty. And the number of babies born live in these two counties actually decreased slightly from 1170 to 1082 in 1966, so that the jump in rates per 1000 births was actually even larger.

There could be little doubt about the statistical significance. Established statistical estimation techniques showed that the possibility of such a fluctuation being accidental was much less than one in 10,000. But this was not all. The students had gathered the data for all five counties surrounding Grundy County, as well as for a control group of six counties as far to the west and north of Grundy as possible within the state of Illinois, counties that bordered neither on the contaminated Illinois River nor on the Mississippi, where the effluent from other nuclear plants upstream in Minnesota and Wisconsin might lead to rises in mortality.

And when we carried out the comparison in the change

of infant mortality rates for these two groups of rural counties of similar climate, medical care, and socio-economic character, the result was even more conclusive: While the mortality rates in the counties around the reactor had increased an average of 48 percent, the upwind control counties actually showed a decline of 2 percent in their average infant mortality rates.

Furthermore, with the prevailing westerly winds, the radio-active gas would drift eastward to Cook County, where Chicago was located, with a population of some five million. Since the radioactivity would have become much diluted with distance, only a small rise in mortality rates of a few percent would be likely. But since so many more children were born every year in Chicago than in Grundy County, the total number of additional deaths would be significant. And when we checked the figures, this is exactly what had taken place: Infant mortality in Cook County had gone up by 1.5 percent, during a time when in New York City it had declined by 6.7 percent.

Since some six million people lived within a radius of 50 to 60 miles from the Dresden reactor, and since the total population of Illinois was ten million, there should have been a significant rise in infant mortality for Illinois as a whole. And there was indeed—from an all-time low point of 23.9 in 1963 to a peak in 1966 of 25.6, in exact coincidence with the peak of gaseous emissions from the Dresden reactor. This was followed by a renewed decline in both recorded gaseous releases and infant mortality as the defective fuel rods were replaced.

With the advice of Dr. Morris DeGroot, head of the Statistics Department, Carnegie-Mellon University, who had become interested in the problem, we applied further statistical tests. The results were always the same: A significant rise and decline in infant mortality in Illinois compared to all other neighboring states in the northern U.S., correlating directly with the rise and decline of radioactive emissions from the Dresden reactor. Relative to Ohio, a few hundred miles to the east, where the infant mortality rate had been the same as in Illinois before the reactor had been started up in 1960, the excess infant deaths in Illinois for the years 1960–68 numbered close to 4000. And for each infant dying in the first year of life, it was well known that there were perhaps three to four that would live with serious genetic defects, crippling congenital malformations, and mental retardation, afflictions in many ways far worse than death in early infancy.

The largest numbers of deaths among the newborn infants were caused by asphyxia or respiratory distress, including hyaline membrane disease, long known to be associated with immaturity, and also general immaturity and "crib death." These were the very causes that had risen sharply all over the world during the period of nuclear testing and had only begun to decline again a few years after the test-ban treaty came into force. Yet here in Illinois, they were still increasing. And among the older infants, noninfectious respiratory disease deaths rose almost 90 percent, and bronchitis almost 50 percent, in the two years after 1964.

In fact, for all ages, there was a rapid rise in deaths due to such lung diseases as emphysema and bronchitis after the onset of the Dresden emissions. The rise was far greater than in more heavily polluted New York. In the ten years between 1949 and 1959, these death rates in Illinois increased by only 9 percent, but they rose by 75 percent in the short period from 1959, when the reactor was completed, to 1966, the last year for which data were available. This was more than eight times the previous annual rate of increase.

Thus, the radioactive gases released from reactor stacks, gases which had been widely regarded as relatively harmless, now appeared to be far more serious in their effects than had been anticipated. Although these gases do not concentrate and remain in the human body, they do dissolve readily in the bloodstream and especially in the fatty parts of many cell membranes when they are inhaled over periods of hours or days. And some of them transform themselves into the biologically damaging cesium, strontium, and yttrium inside the body. As a result, the internal radiation damage to the small air sacs of the lungs, which are lined with cells that produce a crucial fatty substance (lipid) that acts to keep these air sacs open when the air is exhaled, could be far more serious in causing respiratory damage than the external radiation dose from the radioactive gases outside the body.

There was still another way, more indirect but more efficient, in which small amounts of radioactivity could produce deaths from respiratory problems, especially in the newborn. Some of the radioactive chemicals produced by the fission of uranium—such as yttrium 90, the daughter product of strontium—were known to concentrate in the pituitary gland. And recent studies had revealed that the critical lipid needed to prevent the lung from collapsing was produced in special cells of the lung under the chemical control of the pituitary gland in the last few weeks of fetal development just before birth. Thus, even slight damage to the pituitary gland from radioactivity in the air or in the mother's diet could lead to a slight retardation in development, so that the lung would not be quite ready to function properly immediately after birth. And the result would be that otherwise apparently normal babies would be born underweight and would succumb to respiratory failure shortly after birth.

The rise in infant deaths from respiratory diseases associated with immaturity also indicated that the atmospheric reactor releases should be causing an increase in low-birthweight babies. This expectation was confirmed by the data for Grundy County, where the Dresden reactor was located. The number of low-birthweight babies born in this county rose and declined in exact synchronism with the measured gaseous emissions, the rises going as high as 140 percent. No such increases in the number of underweight babies took place in the six control counties more than 40 miles west of the reactor.

The sudden rise in emphysema and bronchitis all over the United States and other countries, noted by I. M. Moriyama, followed the onset of large-scale atmospheric releases of radioactive gas and dust in the early 1950s, also fitted the hypothesis that radioactivity in the air was causing lung damage. When we plotted the emphysema and bronchitis death rates for the states where ordinary air pollution was lowest but radioactivity in the air itself was highest, such as dry, dusty Wyoming, Utah, and New Mexico, where the winds picked up the radioactive dust again and again, we found that after declining in the 1940s, the respiratory death rates per 100,000 people suddenly began to rise sharply between 1946 and 1951, exceeding those in the much more polluted but higher-rainfall states of the east such as New York and Massachusetts, where the radioactivity was cleansed from the atmosphere and soaked into the ground by the rains. By the early 1960s, even in the heavily polluted coal- and steel-producing state of Pennsylvania, these types of respiratory deaths, normally attributed only to ordinary air pollution, were lower than in the clean mountain air of Wyoming.

Clearly, air pollution from ordinary fossil-fuel-burning power plants, which had doubled steadily every ten years for many decades, could therefore not be blamed for all of the alarming rise in lung disease deaths. Instead, all the evidence pointed to radioactive air pollution, both from fallout and from nuclear power plants, as the greatest single contributor to the rise in all types of chronic lung disease around the world, multiplying the effects of the other pollutants—including cigarettes, as in the case of the uranium miners.

Furthermore, there was one source of radioactive pollution that was potentially even more serious than the boiling-water reactors. This was the effluent from the nuclear-fuel processing plants. These plants recovered uranium from the spent reactor fuel elements, as well as plutonium, which could be sold back to the government for use in building bombs and missile warheads. In the process, radioactive gases and large amounts of other fission products were discharged into the air and adjacent rivers. Here all the efforts to prevent the escape of radioactivity from the reactors themselves were therefore completely nullified.

The students gathered the data for the first commercial fuel-reprocessing plant, located in West Valley, New York, some 25 miles south of Buffalo. This plant had gone into operation in April 1966, and reports on its radioactive gas releases, as well as measurements of liquid waste releases and doses from the food produced nearby, had just been published by the U.S. Bureau of Radiological Health in May 1970. When we looked at the available data for infant mortality in Cattaraugus County, where the plant was located, we saw that infant mortality had jumped up 54 percent between 1966 and 1967, far above the rate for New York State as a whole.

Once again, every adjacent county had gone up dramatically at the same time, while the next ring of counties rose only slightly. And those beyond 50 miles all showed declines in their infant death rate, as did New York State as a whole and all the adjoining New England states to the east.

But this situation was not confined to New York State. In some of the counties in Pennsylvania just to the south of the plant, the same rise in infant mortality had taken place. Warren County, directly to the southwest along the valley of the Allegheny River, had gone up almost exactly the same amount as Cattaraugus County. And along the Allegheny River below Cattaraugus County where the plant was located, infant mortality had either jumped up or refused to decline further, the effect diminishing with distance all the way down to Pittsburgh.

The map of the area made the explanation evident: The small tributaries that flowed into the Allegheny River as it passed through New York State originated within a few miles of the fuel-processing plant near West Valley, where the radioactivity discharged from the stack and the storage reservoirs seeped into the watershed for the entire Allegheny River system. Along the Allegheny near the Pennsylvania border, infant mortality had risen 56 percent in Warren County and 48 percent in Venango, through which it passed next. Even as far away as Armstrong County, more than 100 miles downriver, infant mortality had gone up 4 percent that same year, while Pennsylvania as a whole showed no such rise, though it was not declining as rapidly as rural states having no nuclear facilities. Evidently it was not just the inhalation of the atmospheric gases that was important in infant mortality, but also their deposition by rainfall in the headwaters of the Allegheny river, contaminating water, fish, milk, and vegetables with radioactive cesium, strontium, iodine, and other toxic elements.

When we checked the levels of radioactivity in milk reported by the Public Health Service, we found the confirmation of what we had begun to suspect: Of all the milk-sampling stations in the entire United States reporting for the 12-month period ending in March 1970, only those in Pennsylvania showed a level of short-lived iodine 131 greater than 1 micromicrocurie per liter. And there was a more disturbing piece of evidence: strontium 90 had climbed back up to more than half the level that existed at the peak of atmospheric weapons testing, and it was still rising. The West Valley plant was the first commercial facility of its type in the United States, and it emitted far greater amounts of toxic radioactive elements into the environment than did any single nuclear reactor, including Dresden. Yet many more of these plants were planned for the future.

There simply could be no further doubt as to the cause of the rising infant mortality around the West Valley plant: Measurements carried out by the Public Health Service and published in May 1970 showed that, aside from the dose produced by the krypton gases released into the air, doses as high as 250 millirads from cesium and 532 millirads from strontium would be received in a single year by any individuals who ate significant amounts of the area's heavily contaminated fish and deer. These were doses much greater than the 100 millirads normally received from natural background radiation. In fact, they far exceeded even the annual doses during the height of nuclear testing. And these dose calculations were only for the adult, and not for the much more sensitive fetus and infant, where the even more intensive concentration in various critical organs would make the doses far higher still.

The findings were further confirmed when the infant mortality rates in the counties around the AEC's Hanford Laboratories in Washington were graphed. It was at Hanford that nuclear fuel was first processed to produce the plutonium for the Trinity explosion at Alamogordo, New Mexico, in 1945. During this period large quantities of radioactivity were released by the Hanford plant. These releases had explained the early infant mortality rise in Montana and North Dakota that showed up on the Trinity map. But at the time, I did not examine the effect on the counties around the Hanford works itself. Now, when we compared the infant mortality rate for 1945, after the emissions had occurred, with the rate for 1943, before the plant had been started up, we found that the rate for Benton County, where the plant was located, had jumped 160 percent.

Umatilla, the adjoining county to the south, had gone up 60 percent, while Franklin, directly to the east, increased 50 percent and Walla Walla, just to the southeast of Franklin, rose 10 percent. Yet infant mortality for the state of Washington as a whole declined, as it also did in Oregon.

And within a few more months, the results for Consolidated Edison's Indian Point plant on the Hudson River in Westchester County, 20 miles north of New York City, disclosed similar rises and declines in infant mortality that correlated with the rises and falls in the plant's radioactive releases, including observable effects on New York City itself. Yet this was a pressurized-water reactor, the type that generally had the lowest releases of all, and it was located in an area of excellent medical care. Evidently it had not been possible to maintain the standards of a naval-type plant and remain commercially competitive with the boiling-water plants. Similar situations existed around all the reactors we checked in various parts of the country. Even the small research-type reactors, such as the TRIGA, installed on college campuses and in laboratories all over the world, appeared to be capable of causing the same effect. When figures on the year-by-year emissions of the TRIGA reactor at Pennsylvania State College became available, we compared the infant mortality rates in the surrounding town, State College, with those in Lebanon City, a similar town some 100 miles to the east. State College showed precipitous rises and falls in infant mortality, corresponding closely with the rises and falls in emissions from the TRIGA. The State College rate went from 9.9 per 1000 births in 1963 to 24.7 in 1968. During the same period, the rate in Lebanon City, as well as in Pennsylvania as a whole, declined steadily from the peak reached during the atmospheric tests of 1961–62.

Since the population of State College was comparatively small, however, a remote possibility existed that these increases could be due to chance fluctuations. So we next examined infant mortality rates around the TRIGA on the University of Illinois campus in Urbana, where the population was much larger. From 1962, when the reactor commenced operation, through 1965, the year it reached full power, infant mortality increased by 300 percent. In this study, for the first time, we also had an opportunity to look at another category of possible radiation effects: deaths from congenital malformations. During the same period in Urbana these deaths increased by 600 percent, from 3.5 per 100,000 in 1962 to 23.5 in 1965. And in 1968, after the reactor was shut off, they turned downward to 6.6 per 100,000, while infant mortality showed a similar drop. In McLean County, which extended 20 to 60 miles northwest of Urbana and thus would not have been significantly exposed to the effluent, both categories of death declined steadily throughout the same period. The surprising strength of the effects from the TRIGA emissions, which were much lower than the emissions from the larger reactors we studied, could be explained by the fact that the TRIGAs were located right in the middle of densely populated areas. Therefore, the emissions would reach the developing infants in much more concentrated form, with much less time for the short-lived isotopes to lose their radioactivity.

It was the announced intention of the AEC, numerous public utilities, and the government that this country's energy needs would be supplied largely by nuclear-power reactors in the near future. Only fifteen or twenty such reactors were in actual operation, but more than a hundred were under construction or planned, as were the necessary number of fuel-reprocessing plants. But if our findings proved correct, then the entire program, with its phenomenally large investment of funds and scientific energy, would become virtually useless in its present form. Considering the apparent effects from normal operation of these plants, during which no more than one ten-millionth of their stored-up radioactivity had ever been discharged, a single large accidental release could be a national catastrophe of nuclear warfare dimensions. If the general public grasped this fact, then most people would probably consider the risk of this technology far too great to be accepted. But through all the years while reactor technology was being developed, the possible dangers of low-level radiation—either from fallout or from nuclear power plants—had been publicly minimized by the military, by industry, and by the health agencies that had given their stamp of approval to nuclear activities. The warning signs had been ignored or suppressed. And little or no funds had been made available for development of the potentially safer and more efficient alternatives to nuclear power, such as coal gassification or magnetohydrodynamics, which would permit the continued use of the still-enormous reserves of fossil fuels. Little or nothing was done to find means of harnessing the vast stores of geothermal energy in the crust of the earth, or the pollution-free energy of the sun. Yet there was little question that these alternative means of electric power production could have been successfully developed.

Our reactor findings were met with opposition as strenuous as that which greeted the evidence on the effects of fallout. Notable among our critics was Edythalena Tompkins, a public-health scientist who was recently placed in charge of all studies of radiation effects on the population by the U.S. Environmental Protection Agency. Edythalena Tompkins was also the wife of Paul C. Tompkins, the director of the Federal Radiation Council who in 1964 had presided over the first raising of permissible radiation levels in history. Additionally, she had been a critic of the fallout evidence, particularly that relating to the effects of the Trinity test. In the spring of 1970, a student at the Pittsburgh School of Public Health had informed me that Mrs. Tompkins had told him there were serious errors in my map of infant mortality after the Trinity explosion. My map had shown no rise in infant mortality among the white population in three states that were in the path of the fallout—Oklahoma, Florida, and South Carolina. The explanation for this was that according to the official weather map these states had received little or no rain and thus little fallout during the week following the Trinity explosion. This fact had provided an important confirmation of my hypothesis. However, Mrs. Tompkins had told this student that just the opposite had been the case: The infant mortality in these states had actually *risen,* just as it had in the states that received the rains. He then gave me a series of five maps that had been prepared by Mrs. Tompkins, and on all of these maps the three key states did indeed show sharp increases in white infant mortality during the five years following Trinity.

This appeared to be a devastating piece of evidence, but my associates and I rechecked our figures and found that only in these three crucial states did they differ from those on Mrs. Thompkins's maps. I suggested to the student that he himself recalculate the figures. After doing so he informed me that our figures were the correct ones and then called Mrs. Tompkins for an explanation. Mrs. Tompkins said that she had evidently made a mistake, but that in any event these maps were not intended for publication. Subsequently, however, AEC representatives and members of the Joint Committee on Atomic Energy stated publicly that my infant mortality figures for the states that had low rainfalls after the Trinity test had been proved to be inaccurate and that the true figures completely invalidated my conclusions. Yet the only time these particular figures had ever been challenged was by Mrs. Tompkins.

After our group began making the reactor findings public, Mrs. Tompkins, by now with the newly formed Environmental Protection Agency to which she and her husband had been transferred, began conducting her own studies of this subject too. Her method was to calculate the infant mortality rates in a series of circular areas surrounding the reactor, and compare these figures for five-year periods before and after the reactor had gone into operation. She concluded that even the most heavily emitting boiling-water reactors had no detectable effect on infant mortality.

Since vital statistics are recorded by county and not by circular regions around reactors, however, Mrs. Thompkins's method first of all necessitated that she make her own estimates of both the population figures and the infant mortality rates. And her use of concentric ring-shaped areas omitted a very important consideration: The emissions from reactors are not evenly distributed in all directions. Their distribution depends not only on the direction of the prevailing winds, or on geographical features such as high mountains and resulting differences in rainfall, but also on the discretion of the reactor engineers, who can and do time the releases to coincide with a certain wind direction that may or may not be the prevalent one. And then, of course, the counties that take their drinking water and fish from the rivers, lakes, or oceans into which the reactor releases its liquid effluent would also be expected to show sharper increases than the others. Thus, highly asymmetrical situations can develop around reactors, situations in which the counties most heavily exposed to the effluent show sharp rises and falls in infant mortality that correlate directly with rises and falls in the reactor releases, while in other counties, such as those upwind to the west of Dresden, the rate may continue the decline that began shortly after the cessation of atmospheric testing. Thus, if all the surrounding counties are averaged together over five-year periods, as in Mrs. Tompkins's method, the overall figure may show little or no increase in infant mortality. Furthermore, in the case of reactors that began operation in the early 1960s in areas that had received heavy fallout (as had the three boiling-water reactors studied by Mrs. Tompkins), it is possible by this technique to demonstrate an actual *decline* in infant mortality after the reactors were started up and the fallout levels died down. But in all of these cases, if one examines the yearly figures, the infant mortality rates in the counties heavily exposed to the reactor effluent show sharp rises and falls in direct correlation with the releases, declining steadily with distance in any direction from the reactor when the counties are of similar socio-economic and climatic character.

Significantly, an independent statistical study of this subject was presented at a scientific meeting in July 1971 by Dr. Morris H. DeGroot, head of the Department of Mathematical Statistics at Carnegie-Mellon University. Dr. DeGroot found that infant mortality increases did take place in close correlation with releases of radioactivity from the heavily emitting reactors at Dresden, Illinois; Indian Point, New York; and Brookhaven, Long Island. Perhaps most important was his finding that in the area around the reactor at Shippingport, Pennsylvania—the only other reactor studied by Dr. DeGroot—there was no correlation between releases and changes in infant mortality. As the official release figures showed, the Shippingport reactor had the lowest gaseous emissions of any reactor in the country, since it was a non-commercial naval submarine type of plant.

But later in 1971, the most comprehensive independent

study of all was completed. It was conducted by Dr. Lester B. Lave and his associates, Dr. Samuel Leinhardt and Martin B. Kaye, of the Graduate School of Business Administration at Carnegie-Mellon University. This was a study of fallout effects, but the results apply equally to reactor emissions. The three scientists concluded that, during the time period studied (1961–67), fallout appears to have been the single most important factor affecting fetal, infant, and adult mortality, more important than ordinary air pollution. Through the use of computerized statistical techniques they corrected their estimates to account for the effects of such variables as sulfur dioxide, socio-economic factors, background radiation, and others in 61 metropolitan areas of the United States. The principal findings and their implications may be briefly summarized as follows:

Infant mortality is strongly associated with levels of strontium 90 and cesium 137 in milk, especially the former. The association is such that for every single micromicrocurie of strontium 90 per liter of milk there is an increase of 12 infant deaths per 100,000 births. Since, during 1961–67, there was an average of 15.8 micromicrocuries per liter of milk in the U.S., then these findings indicate that during this period there were close to 7600 infant deaths *every year* due to fallout. For the world population, this would mean an extra 100,000 infant deaths per year. But during the peak of testing, these levels reached between 50 and 100 micromicrocuries per liter in many locations around the world, and as late as 1971 they were still between 5 and 15 in most parts of the northern hemisphere. And they then began to rise again following the large French and Chinese test series and the rapid growth in releases from nuclear reactors and fuel reprocessing plants.

Dr. Lave's group also found that mortality rates for the whole population—in other words, all causes of death among all ages—were also highly correlated with fallout levels. The calculations showed that there were 1.29 extra deaths per 100,000 people for each single micromicrocurie of strontium 90 per liter of milk. At the 1961–67 levels, this amounts to some 40,000 extra deaths each year in the United States, and thus some 600,000 among the world's population of over three billion people.

And during the fifteen-year period of heavy nuclear testing that began in the early 1950s, when the short-lived iodine and other isotopes were added to the strontium 90 in the milk, there would have been many millions of extra deaths.

At long last, more than a quarter century after Hiroshima, studies of the health effects of fallout were being made by independent scientists outside the government such as Lave, Leinhardt and Kaye. But as I was not to learn until much later, neither the public nor the scientific community at large would be able to learn of these results. When the Carnegie-Mellon scientists submitted their paper to *Science,* Abelson refused to publish it, even though a similar paper by the same group linking ordinary air pollution to mortality increases using the same statistical techniques had been published by *Science* earlier.

The paper was finally accepted for publication in the much less widely read journal *Radiation Data and Reports,* published monthly by the Environmental Protection Agency. But the important findings of Lave, Leinhardt and Kaye never appeared in print. Just before publication, when the plates had already been prepared, the authors received word from the editor that objections from highly placed government officials forced them to destroy the plates. The article has never appeared in the scientific literature, and at the end of 1974, publication of *Radiation Data and Reports* ceased with the December issue after fifteen years of providing the only comprehensive source of data on radioactivity in the environment, following deep budget cuts in the Office of Radiation Programs ordered by the Nixon administration.

15

Fallout at Shippingport

THE STUDIES of Lave and DeGroot provided independent evidence that infant mortality was correlated with low-level radioactivity from nuclear-weapons fallout and reactor releases, but a number of puzzling questions remained unanswered. It was understandable in the light of Dr. Stewart's latest findings, published in 1970, that infant mortality might go up significantly as a result of early intrauterine exposure due to the hundredfold greater sensitivity of the fetus in the first three months of development as compared to the adult. It was difficult to understand, however, how total mortality rates, dominated by the older age groups rather than by the small number of newborn infants, could possibly be affected as strongly as Lave's study had shown.

Still another puzzle was the finding by DeGroot that although infant mortality rates in Beaver County, where the Shippingport reactor was located, did not decline as rapidly as for the state of Pennsylvania as a whole, there was no correlation between the abnormally high infant mortality rates and the officially announced small releases from the plant.

Both of these puzzles were destined to find their solution in a most unexpected manner within a year after DeGroot's and Lave's studies had been completed. Late in 1972, a notice in the Pittsburgh newspapers announced that hearings would shortly be held by the Atomic Energy Commission to grant an operating license for the Beaver Valley Unit I reactor, which was then nearing completion. This power station was being built right next to the original Shippingport reactor on the Ohio River, some 25 miles downstream and to the west of Pittsburgh. According to the newspaper story, it would be of the same pressurized-water type that had been pioneered in Pittsburgh by Westinghouse, under Admiral Rickover's direction, except that it would be some ten times larger.

Knowing that it was a naval type of reactor with a double cooling loop to minimize the amount of gas that would have to be discharged into the atmosphere caused me to feel little concern, especially in view of the fact that the AEC had only recently announced that it was proposing to tighten up the standards for permissible emissions. (These new standards had been issued following hearings in Washington at which I had been asked to testify in behalf of various environmental groups on the need to lower permissible doses.) Also, Westinghouse had just announced that it had been possible to operate Shippingport with "zero" gaseous releases in 1971, so that I felt certain that this much more advanced new power station only a short distance upwind from Westinghouse headquarters and the Bettis Nuclear Laboratories, where the first submarine reactors had been built, would surely be provided with the very latest in the available equipment for containing all radioactive gases.

Thus, when some of my students asked me whether I planned to attend the hearings I expressed no great concern, saying only that I might take a look at the Safety Analysis Report being kept in the public library of the nearby town of Beaver, a few miles from Shippingport, to make sure that the planned emissions were indeed as low as I expected them to be.

A few weeks later, an opportunity presented itself to check on the proposed releases. I had to go to the nearby Pittsburgh airport to pick up my mother, and since the Beaver County Library was only a few miles from the airport, I left a few hours early to check the figures.

Since I had examined similar reports for the Davis-Besse and other plants within the past year, it did not take me long to find the information I was looking for. But what I found shocked me profoundly. Instead of gaseous releases of only a small fraction of a curie, such as had been reported for Shippingport in recent years, the more advanced commercial plant about to go into operation was apparently designed to release some 60,000 curies of fission gases per year into the already heavily polluted air of the Ohio River valley. This was millions of times more than was claimed to have been discharged annually from the old Shippingport plant in recent years, even though the power output would be only ten times greater.

In fact the summary of past releases from nuclear facilities published by the Bureau of Radiological Health had listed only 0.35 curies of fission gases at the time of the highest reported discharges back in 1963, for which the calculated dose was 0.87 percent of the maximum permissible of 500 millirems to someone living near the plant. This meant that the estimated radiation dose produced by 0.35 curies was only about 4 millirems. Yet even at these relatively low calculated external doses (due to gas releases), there seemed to be a disturbing rise in infant mortality in surrounding Beaver County and especially the nearby town of Aliquippa, some 10 miles to the east in the Ohio valley.

There were thus only two possibilities. If the reported figures on the likely magnitude of gaseous releases from the new large reactor were correct, there would very likely be a major increase in infant mortality and other detrimental health effects unless vastly more efficient means of trapping the gases were installed to bring them down to the levels reported for the existing reactor.

The other possibility was that the actual releases from the Shippingport plant had somehow been much larger than the amounts officially reported. And this would of course explain why DeGroot did not find a relationship between the tabulated releases and the yearly changes in infant mortality for the Shippingport plant.

Deeply troubled by these findings, I decided to contact the utility lawyer for the City of Pittsburgh, Albert Brandon, who had long been battling the Duquesne Light Company's growing requests for rate increases needed to finance the escalating cost of the Beaver Valley nuclear plant. My hope was to persuade the city to intervene in the upcoming license hearings in order to get to the bottom of the disturbing discrepancy between the annual claim for "zero-release" nuclear plants and the actually planned emissions. Even though it was too late to stop the plant from going into operation, perhaps it would still be possible to force the utility to install the latest equipment for trapping the radioactive gases so as to reduce to a minimum the health risk to the people living in the area.

Concerned by these facts, Brandon promised to discuss the matter with the mayor, Pete Flaherty. A few days later, a meeting was arranged, and after a brief discussion, Flaherty agreed to have the City of Pittsburgh become an intervenor in the upcoming license hearings, together with a group of local environmentalists to whom I had previously outlined my findings.

Shortly after the public announcement that the City of Pittsburgh would intervene in the hearings for the new plant, I received a telephone call from a man who identified himself as the manager of the new power station being built at the Shippingport site. He said that great efforts were being made to assure the safety of the people in the area, and that he would be glad to send me the detailed plans for the environmental monitoring that would be done to assure that no harmful amounts of radioactivity could reach the public.

Within a day, a large manila envelope was delivered to my office at the university from the Duquesne Light Company. As I leafed through its contents, I noticed a series of documents entitled "Pre-Operational Environmental Radioactivity Monitoring Program at the Beaver Valley Power Station" in the form of quarterly reports for the years 1971 and 1972. The documents had been prepared by the N.U.S. Corporation of Rockville, Maryland. These were apparently part of the Environmental Report for the Beaver Valley Power Station Unit II Construction Permit Application, submitted to the AEC in November 1972 as required by the new National Environmental Protection Act, which had just come into effect. Thus, the data were gathered to establish the radiation levels existing at the site prior to the operation of the new plant, providing a baseline for comparison with later measurements that would be gathered once the plant had gone into operation.

As I began to look through the tables with their long lists of numbers, I noticed that there were some very high measurements for the external gamma doses in early 1971, measured in microrems per hour. When I worked it out in the more familiar units of millirems per year, I could hardly believe the result: In March the rate was 370 millirems per year for Station No. 10, located in the town of Shippingport, compared to the normal values for the area of 70 to 90 millirems per year. There were a few more readings at this location in the range of 300 to 350 millirems per year by June, and not until January of 1972 did the numbers return to the normal rate of 86 millirems per year.

Other locations showed comparable peaks of gamma radiation, but the highest were in the town of Shippingport closest to the site or on the site itself. Could it be that these extremely high radiation dose rates were produced by the old Shippingport plant, for which the official reports had shown almost no gaseous releases at all?

Turning to the tabulations of strontium 90 in the milk, I saw immediately that the levels measured in the farms around Shippingport were much higher than in Pittsburgh, Harrisburg, Cincinnati, and Buffalo as reported in "Radiation Health Data and Reports" for the early part of 1971. The fact that the extremely high readings were confined to the Shippingport area made it unlikely that they were due to worldwide fallout from high-altitude atmospheric bomb testing.

To check this further, I plotted the concentrations of strontium 90 in the soil and found that it dropped off sharply with distance away from the plant both east to west and north to south. In April of 1971, the levels within three-quarters of a mile were fifty times greater than the typical levels produced by worldwide fallout, and by early in 1972, the rains had apparently washed most of the activity into the Ohio River, the measured levels having gone down from their peak of 6000 picocuries per kilogram to less than 100.

Clearly, such a highly localized concentration of strontium 90 in the soil centered on the Shippingport plant could not be explained by worldwide fallout, which is more or less uniformly distributed around the globe as the rains bring down the fine particles circulating in the upper atmosphere.

Still further confirmation of the localized nature of the radioactive contamination came from the measurements of short-lived iodine 131 in the milk. Beginning in December of 1971 and peaking in February 1972, the levels of iodine for the six dairies within a 10-mile radius started to rise above 10 picocuries per liter, the Range I reporting level set by the Federal Radiation Council for continuous consumption, reaching as high as 120 picocuries per liter. This was well above the 100 picocurie-per-liter limit of Range II, and it equaled the kind of values reached in the eastern United States during the height of nuclear-bomb testing.

Yet when I looked up the monthly iodine 131 levels for other locations in Pennsylvania (such as Erie, Harrisburg, and Philadelphia) in "Radiation Health Data and Reports," they were all listed with "zero" values, or below the limit of detection. Clearly, it was extremely unlikely that any Chinese fallout would somehow concentrate radioactive iodine 131 over the Shippingport site, leaving the nearby areas of Ohio and Pennsylvania without any detectable increases of radiation in the milk.

As a final check, I compared the monthly values of strontium 90 in the milk within a 10-mile radius of Shippingport with the monthly electrical power output in kilowatt-hours published in *Nucleonics Week*. Both strontium 90 and power output peaked in January 1971 and again in April, moving up and down together until the plant was closed for repairs later in the summer. After the plant was shut down, both the local and the Pittsburgh milk showed a sharp reduction in strontium 90 levels, from a peak of 27 picocuries per liter nearest the plant in early 1971 down to 7 picocuries per liter measured in Harrisburg that summer. As I learned later from an analysis of the milk-marketing reports, the city of Pittsburgh obtained about a third of its milk from an area within 25 miles of the Shippingport plant. This finding was consistent with the fact that the Pittsburgh milk showed strontium 90 concentrations some 30 percent higher than the Cincinnati and Philadelphia milk in early 1971.

Yet during the time of the sharp peaks in radiation levels in the air, the soil, and the milk that occurred between January and June of 1971 near Shippingport, there were no nuclear-weapons tests carried out in the atmosphere by any nation as reported in the monthly issues of *Radiation Health Data and Reports.*

After weeks of graphing and analyzing the data with the help of colleagues, volunteers from local environmental organizations, and students at the university, there could be no doubt about the result: The data collected by the Duquesne Light Company's own hired team of experienced health physicists clearly indicated that the Shippingport plant must have been the source of radioactivity in the environment many thousands of times as great as had been claimed in the official reports to state and federal agencies. Instead of annual radiation doses of less than 0.5 millirems claimed by the utility, the combination of external radiation (measured by the dosimeters) and internal radiation (from the gases that were inhaled or ingested with the milk, the water, and the local meat and vegetables) was many hundreds of millirems per year. Indeed, this dosage exceeded the level of radiation that was received by the people of this area during the height of nuclear-weapons testing. Moreover, the scientists who had carried out these measurements had clearly failed to warn either the utility officials who had hired them, the public-health officials at the state or federal level, or the public, whose health and safety were being endangered by the secret fallout from the plant.

Faced with these disturbing discoveries, the leaders of the local environmental groups in Beaver County decided to hold a public meeting at which both the Duquesne Light Company and spokesmen for a Pittsburgh environmental group would be able to present their views to the people of the area. The meeting took place early in January of 1973 at a shopping

mall in the town of Monaca, just a few miles from the Shippingport plant. After the superintendent of the Shippingport plant
explained that the new power station would be "the Cadillac
of the industry"—with a waste-disposal system that would permit only "minimal" amounts of radioactivity to escape—the
head of Environment Pittsburgh, David Marshall, and I presented the data gathered by the Duquesne Light Company's
own consultants. Slide after slide showed the localized concentrations of radioactivity in the milk, the soil, and the river
sediments rising to many times their normal value, together
with the peaks during the months when there was no nuclear-
weapons testing. Obviously, the findings in our presentation
were completely at variance with what the utility had told
the local people over the years.

The Duquesne Light officials were unprepared for this damaging evidence and could only lamely repeat their assurances
that the new plant would have negligible impact on the health
of the public. It took them a few days to prepare an advertise-
ment for the *Pittsburgh Post-Gazette* in which they claimed
that they had operated their Shippingport facility safely—
without releasing more than a small percentage of the releases
allowed by the Atomic Energy Commission and the Commonwealth of Pennsylvania, and therefore without injuring
any member of the public. But the people who had attended
the meeting were no longer so certain that this was the case,
and there was a demand for an independent investigation of
these disturbing findings by the various environmental groups
in Pittsburgh and Beaver County before a new and still larger
reactor would be given a license. This demand was supported
by the mayor of Pittsburgh, Pete Flaherty, and his utility lawyer, Albert Brandon.

Confronted with the evidence of very high levels of strontium 90, cesium 137, and iodine 131 in the area in 1971, while
"zero" release had been officially reported, I began to wonder
about earlier releases. The plant had been in operation since
1958, so in light of the unreliable claims by the company, I
wondered if there might indeed have been long-term exposure
to the people of Beaver County and nearby Allegheny County,

in which the city of Pittsburgh was located. In particular, enough time had elapsed for leukemia and cancer to develop, so that one might for the first time be able to determine whether the operation of commercial nuclear plants did or did not lead to the same kind of cancer increases that I had begun to see following the start of nuclear-weapons tests in Nevada, the Pacific, and Siberia.

My students and I started to examine the annual vital statistics reports for Beaver County, Allegheny County, and the major towns at different distances from Shippingport up and down the Ohio River. Within a few days the first results were tabulated, and the figures were startling. In the town of Midland, just a mile downstream from Shippingport, the people drank the Ohio River water. The cancer death rate in this town had risen from a low of 149.6 per hundred thousand population in 1958, when the plant started to operate, to a peak of 426.3 by 1970. This was an increase of 184 percent in only twelve years.

For Beaver County as a whole, surrounding the plant, the rate had risen from 147.7 to 204.7 in the ten years from the time the plant had gone on line with so much hope for a cleaner and healthier environment. This was a rise of close to 40 percent during a time when the state of Pennsylvania as a whole showed an increase of only 10 percent and the U.S. cancer mortality rose by only 8 percent. From a low of 293 cancer deaths in Beaver County in 1958, the number had risen to 418 by 1968, an increase of 115 cancer deaths per year, when there should have been no more than an additional 30 if the county had continued to follow the average pattern for the state.

Likewise, the Pittsburgh cancer death rate had climbed by 31 percent between 1958 and 1968, despite the steady cleanup of ordinary air and water pollution that had begun right after World War II, when the burning of soft coal in the city was ended and a major effort was begun to clean up the air and water.

Similarly, in the towns along the Ohio River downstream from Shippingport and Midland, cancer rates had climbed sharply, the more so the closer they were to the plant. For

East Liverpool, just across the border in Ohio and some 10 miles downstream, the cancer death rate had risen 40 percent by 1968 and 67 percent by 1971. In Steubenville, some 30 miles downstream, the cancer mortality rate was up 25 percent by 1968, and even as far away as Cincinnati, some 300 miles down the Ohio River, the cancer deaths had climbed 24 percent, while they increased only 6 percent for Ohio as a whole.

Further evidence suggested that the releases from Shipping-port had added heavily to all the other sources of carcinogens, from bomb tests to chemical plants. The city of Columbus, Ohio, which did not use the Ohio River for its drinking-water supply, actually experienced a 10 percent decline in its cancer rate during the same period, even though it suffered from all the other likely sources of carcinogens, including automobile exhaust, cigarettes, food additives, hair dyes, artificial sweeten-ers, and so on.

But if Shippingport was responsible for these striking cancer rises in the towns using the Ohio River for their water supply, then the discharges into the river would have had to be vastly greater than the amounts for which the plant had been licensed. Was there any evidence that the activity in the water had been much greater downstream than upstream of the plant? After all, it was clear that it could not be the milk that was responsible for transmitting the radioactivity all the way to Steubenville and Cincinnati.

Fortunately, there was a way to check this. For many years, the Pennsylvania State Department of Environmental Re-sources in Harrisburg had been making quarterly measurements of the radioactivity in all the major streams of the state at various points along each river. When the students had collected the data for the Ohio and other streams in western Pennsylva-nia, the answer began to emerge. There was a large peak in the Ohio River radioactivity in late 1970 and early 1971, exactly the time when the N.U.S. data had shown a large peak of radioactivity in soil, milk, river sediment, and fish. At Midland, just a little over a mile below the Shippingport plant, the gross beta activity had climbed from a low of only 3 picocuries per liter to a high of 18. But for the two rivers that joined in

Pittsburgh to form the Ohio, the Allegheny and the Monongahela, measured at locations more than 30 miles away, upstream to the east the rise was no greater than 5 of these units.

Thus, the rise in river radioactivity could not have been due to fallout, which would have affected the more distant upstream areas just as strongly. But it was consistent with high, unreported gaseous releases that would settle on the land and then be washed into the Ohio River with the rain and melting snow. In fact, the rapid disappearance of the high values of long-lived strontium 90 in the soil around the Shippingport plant between early 1971 and 1972 could be explained only by the action of rain carrying the radioactivity from gaseous releases into the local streams and rivers. This possibility was further supported by the fact that the two nearest small rivers that joined the Ohio just a few miles *upstream* from Shippingport, the Beaver River and Raccoon Creek, both showed even larger rises in activity, reaching peaks of 20 picocuries per liter during the same quarter.

It was apparently not any direct liquid discharges that were involved, which by the terms of the original license were to be held to less than 0.56 curies. Rather, the radioactivity must have originated from airborne releases that settled on the surrounding land as far upstream as 20 to 30 miles. Only releases into the air could also explain the large increases in milk activity all around the farms surrounding the Ohio River in Beaver County.

This would make it possible to understand the paradoxical finding that even "upstream" locations and tributaries of the Ohio within 20 to 30 miles, showed peaks in radioactivity when the local milk rose in strontium 90, cesium 137, and iodine 131. And it would explain why cancer rates in cities as far away as Pittsburgh, upstream by 25 miles, could have their water supplies contaminated. The wind was blowing the radioactive gases up the Ohio Valley to the streams that filled the reservoirs serving Pittsburgh, just as the fallout from the "Simon" shot in Nevada had contaminated the reservoirs of Albany and Troy back in the spring of 1953.

Clearly, if such releases were taking place but were somehow not reported, even cities using tributaries of the Ohio entering the river 10 to 30 miles upstream from Shippingport, as well as communities far downstream, could have their drinking water affected and their cancer rates increased by the invisible, tasteless, and odorless radioactive fallout secretly discharged into the ambient air.

By looking up the amount of water carried by the Ohio per second at Midland for each month of the year, it was possible to calculate how many curies had been carried downstream from the airborne releases in late 1970 above and beyond the amounts in the Allegheny and Monongahela Rivers that joined to form the Ohio some 25 miles upstream from the plant. The total worked out to 183 more curies in the Ohio below the plant in a year than were carried by the Allegheny and Monongahela Rivers, which combined to form the Ohio. This was 300 times more than the original permit had allowed for direct discharges into the Ohio River from the Shippingport plant, and 2500 times more than the 0.07 curies that the Duquesne Light Company had officially reported for liquid discharges in 1970 to the state and federal health agencies.

There were apparently hundreds to thousands of times as many curies of highly toxic radioactivity in the Ohio River than were allowed by state and federal limits, designed to protect the health of the people using the Ohio for their drinking water. The radioactivity did not come from the direct liquid discharges, however, but through the run-off of unreported gaseous releases that had settled on the land.

Here, then, was at least one piece in the puzzle as to why not only infant mortality but mortality at all ages had been affected so strongly, despite the relatively small external radiation doses from gamma rays on the ground that irradiate the whole body uniformly. It was the airborne gaseous activity and the run-off into the rivers serving as drinking-water supplies that had apparently carried the more damaging short-lived beta-ray-emitting chemicals rapidly into the critical organs of the people, in addition to the other pathways via the milk, the vegetables, the fruits, the fish, and the meat that were most important for the long-lived strontium 90 and cesium 137. And although adults were more resistant to the biological damage than the developing fetus, they received the doses steadily over many years rather than just for a few months, by continuously drinking the water, inhaling the gases, and eating the food that was contaminated first by the fallout from the bomb tests, and then by the secret gaseous releases from the peaceful nuclear reactors along the rivers of the nation.

Of equal significance were the implications for one of the most important questions DeGroot was unable to answer: Why had he not found a correlation between the changes in infant mortality in Beaver County and the published radioactive releases in the case of the Shippingport reactor, while he had discovered such a correlation for the other three nuclear reactors he had studied? Clearly, if there existed such large unreported releases as the data gathered by the N.U.S. Corporation, the Environmental Protection Agency, and the State of Pennsylvania seemed to indicate, then one could not possibly expect to find a direct relationship between the *announced* annual releases and the changes in mortality rates.

Now a new and most disturbing question had arisen: How was it possible for large quantities of radioactive gases to escape from the Shippingport plant without being officially reported as required by the existing regulations? Not until many months later was this riddle destined to be solved in a most unexpected manner.

In the meantime, there was a growing public debate over the abnormally high levels of radioactivity around the Shippingport plant and the sharp rise in infant mortality in such nearby towns as Aliquippa. I documented my findings in a report and sent it to the governor of Pennsylvania, Milton Shapp, in January of 1973. Early in the spring, Governor Shapp announced his intention to appoint a special fact-finding commission of independent scientists and public health experts who would hold hearings on the question and issue their own report within a few months.

The latest numbers for infant mortality in Aliquippa, some 10 miles downwind and to the east of the plant, were indeed alarming. For the years 1970 and 1971, the years of high levels of radioactivity, Aliquippa's infant mortality rate climbed to a twenty-year high of 44.2 and 39.7 per 1000 live births. These were more than double the overall state rates of 19.9 and 18.2. Yet back in 1949 and 1952, when ordinary air pollution from the steel mills was much greater, but before Shippingport had started, Aliquippa's infant mortality rates had been as low as 16.0 per 1000 births.

This could not be simply explained by a change in the composition of the population, which had remained essentially constant, the nonwhite population representing 21 percent of the total in 1960 and 22 percent in 1970. And for the State of Pennsylvania and the United States as a whole, infant mortality had resumed its previous decline after the end of atmospheric bomb tests by the United States and the Soviet Union for both the white and nonwhite population.

News of the controversy had reached the cities along the Ohio below Shippingport, and in April I was asked to present my findings at a public lecture at the University of Cincinnati by a local environmental group and university professors concerned about the construction of the Zimmer nuclear power station upstream from the city's water intake. At the end of my presentation, members of the university's Department of Chemical and Nuclear Engineering attacked my findings, charging that numerous state and federal government health agencies, including those of the State of Pennsylvania, had found no substance to my allegations in the past and that I had been repudiated especially by such prestigious organizations as the Health Physics Society, the American Academy of Pediatrics, the National Academy of Science, the Atomic Energy Commission, and the Environmental Protection Agency.

As Dr. Bernd Kohn, director of the Radio Chemistry and Nuclear Engineering Research Center put it: "In each case, an epidemiologist has refuted his claim by the same data." But Dr. Kohn and the other engineers present were unable to point out how else to explain the startlingly high localized values of strontium 90, cesium 137, and iodine 131 in the environment around Shippingport, other than that it was likely to be Chinese fallout.

However, when I showed the data to the mayor of Cincinnati, Theodore M. Barry, he wrote a letter to Governor John J. Gilligan of Ohio, requesting an investigation by the Ohio Environmental Protection Agency. Also, the chairman of the energy conservation committee of the Cincinnati Environmental Task Force, after seeing the data on radioactivity and cancer mortality changes around Shippingport and the other reactors that had been studied by DeGroot and me, announced that he would recommend that the City of Cincinnati become an intervenor in the public hearings on an operating license for the Zimmer plant.

The next day, the *Cincinnati Inquirer* carried the following two headlined stories on its front page: "Mitchell Denies Knowledge of Plans to Bug Watergate" and, just below, "AEC Denies Radiation Damage to Ohio River."

In the light of the enormous discrepancy between the official claims of "zero releases" and the N.U.S. findings of much larger than normal amounts of strontium 90 in the soil, the milk, and the river sediment around Shippingport, the coincidental juxtaposition of these two stories took on an ominous ring. The facts that had emerged so far were hardly consonant with the AEC's claim in the *Inquirer* story that "the release of effluents from the Shippingport Atomic Power Station is carefully controlled and monitored so as not to endanger the public."

The story went on to say that "the radiation levels in these effluents are so extremely low that they pose no threat to the people in the cities mentioned by Dr. Sternglass." It all sounded exactly like the old reassurances that had been issued by the AEC at the time of the nuclear tests in Nevada, and the denial by former Attorney General John M. Mitchell before a federal grand jury that he had any prior knowledge of the Watergate case and always vetoed any bugging plans that were suggested while he was President Nixon's campaign manager.

There would soon be another kind of grand jury appointed to hear the differing claims of government officials and independent scientists who had stumbled upon information that was not meant to reach the ordinary citizen of our country.

Newspaper stories in the Pittsburgh area repeating the denial of large discharges from Shippingport and blaming the high readings either on fallout or on errors in the measurements were clearly indications of deep concern by the AEC, Duquesne Light, and N.U.S. All three organizations now knew that before long they would be facing hearings by an independent body of knowledgeable scientists. The bureaucrats and scientists in the AEC knew that this time the hearings would not be under their control, unlike the case of the usual licensing hearings, where both the hearing officers and the staff were appointed by the agency whose mandated task it was both to promote and regulate the safety of the nuclear industry.

But the full extent of the behind-the-scenes efforts to make the public believe that nothing had happened at Shippingport did not emerge until long after the hearings of the fact-finding commission had taken place at the end of July. The story was pieced together later in an article by a free-lance investigative writer, Joel Griffiths, and published in an article in the *Beaver County Times* on June 7, 1974, after the AEC had issued licenses for the operation and construction of the Beaver Valley Power Station Units I and II.

Quite unexpectedly, the story came to light as the result of a routine request submitted by the attorney for the City of Pittsburgh, Albert Brandon, in connection with the discovery procedures preceding the licensing hearings for the new reactors at Shippingport. (This was a few months after the Shapp Commission hearings in Aliquippa had taken place.) Brandon had asked for copies of all correspondence and internal memoranda connected with the Shippingport controversy in the files of the AEC. And then, one day in the fall of 1973, not long before the licensing hearings were scheduled to begin, a large envelope arrived at Brandon's office with a devastating series of internal memoranda, letters, and other documents revealing what had taken place behind the scenes.

As Griffiths described it in his article, early in 1973 the AEC's Earth Science Branch had conducted an in-depth investigation of the situation and concluded that "it is highly unlikely that the radioactivity was of Chinese origin. Most likely it was either of local origin, or the result of inadequate sampling procedures." Griffiths wrote that this was a crucial finding.

"Local origin" was a euphemism for Shippingport, since there was nothing else in the vicinity that could have produced that amount of radioactivity. Thus, if the radioactivity had in fact been there, Shippingport was clearly implicated. The only other possibility was that maybe the radioactivity had really not been there in the first place.

As Griffiths put it:

> This was where "inadequate sampling procedures" came in. The idea was that N.U.S. might have bungled procedures it had used to measure the radioactivity in the samples of soil, milk, and other items from Beaver Valley and somehow produced hundreds of erroneous readings, and all of them too high. This, however, was synonymous with the conclusion that N.U.S. was incompetent.
>
> There was only one way this question could be settled in a conclusive manner. Some of the radioactivity in the samples that N.U.S. scientists had collected in 1970 and 1971 was long lasting. If N.U.S. could turn up some of the original samples that had shown the high levels, they could be reanalyzed to see if the radioactivity had really been there.
>
> According to the records, N.U.S. conducted a search in February 1973 at its Rockville headquarters to see if any of the original high samples were still around. Unfortunately, it was the company's stated policy not to retain samples for more than a year after analysis, and none could be located.

Griffiths went on to relate an interesting development:

> By this time, a sharp divergence of opinion had grown between N.U.S. on the one hand and the AEC and health agencies on the other. Faced with a choice between attributing the radioactivity to Shippingport or to N.U.S.'s incompetence, the AEC and others picked incompetence and began leveling various technical charges against the N.U.S. reports. This placed N.U.S. in a delicate position. If their reputation was to be salvaged without crucifying their employer, the Duquesne Light Company and the AEC, N.U.S. had somehow to prove that the radioactivity had been there

but had not come from Shippingport. So despite all the evidence, N.U.S. picked fallout.

In March, 1973, N.U.S. completed a draft report on the Shippingport situation, defending the accuracy of its original high readings but attempting to prove that they were not particularly unusual and were probably due largely to Chinese bomb tests.

This draft report was sent to Dr. John Harley, director of the AEC's Health and Safety Laboratory. Dr. Harley had been playing a leading role in the AEC's investigation of the Shippingport affair, and he was well aware that the high radiation levels could not be explained by fallout.

In fact, I knew that he had worked in this field for years and had previously been involved with minimizing the health impact of the fallout from the "Simon" test that had rained over Albany and Troy back in 1953. He had also played a major role in trying to discredit the findings I had made that showed a connection between the upward changes in infant mortality from the atmospheric tests in the Pacific and Nevada and the levels of fallout in the milk and diet through the use of the misleading "gummed film" data, which falsely showed high strontium 90 levels in the dusty, dry areas where the milk levels were actually quite low.

As Griffiths's story indicated:

The memoranda in the AEC files showed very clearly that Dr. Harley was not happy with N.U.S.'s draft report.

In comments for the AEC's files, dated March 8, 1973, Harley fumed: "This draft proves to my satisfaction that the work of this organization is incompetent. . . . It is obvious that their staff is not familiar with the field and is not competent to evaluate their data or those of others."

Harley went on to list several examples of N.U.S.'s incompetence in their attempt to prove the fallout theory and in other aspects of their report, remarking that "Investigation would certainly turn up gross calculation errors or even that some doctoring of the numbers had occurred."

He signed off: "I believe the situation is very serious."

Serious indeed. Could Dr. Harley have been referring to that team of "outstanding scientists" who, according to Duquesne's ads, were engaged in the vital work of making people aware that their large nuclear plant was to be "absolutely safe to the public health"?

Yes, he was.

More serious was that N.U.S. had already performed extensive safety studies for some thirty-four other nuclear power plants, many of which had already started operating.

If they were bunglers . . . .

Dr. Harley's accusations of incompetence were more incongruous in view of the apparent excellent credentials of the N.U.S. staff, including the two members who prepared the draft report.

One, the vice-president in charge of all N.U.S. nuclear safety work, Dr. Morton Goldman, had spent ten years as a nuclear safety expert with the U.S. Public Health Service (now the Environmental Protection Agency) and was a consultant to state and federal health agencies.

The other, Joseph DiNunno, the scientist directly responsible for the Beaver Valley survey, had received all his training and experience in the AEC's own reactor safety branch.

Why, N.U.S. almost was the AEC and EPA. Incompetence? Doctoring of figures?

Nevertheless, a couple of months after Dr. Harley's outburst, the AEC issued a definitive report stating that the high radiation levels had been due to N.U.S. bungling. The report was hand carried to the Pittsburgh newspapers before N.U.S. even got a chance to look at it.

Shortly thereafter, on June 7, 1973, according to AEC documents, the president of N.U.S., Charles Jones, called the AEC. Jones maintained stoutly that the radioactivity really had been there and that there was nothing wrong with N.U.S.'s methodology.

The AEC representative to whom he spoke, Dr. Martin B. Biles, director of the Division of Operational Safety, disagreed. Jones than complained that the unfavorable publicity was damaging his company and something must be done. Dr. Biles suggested a meeting.

On June 20, 1973, a meeting was held between Dr. Goldman and DiNunno of N.U.S., Dr. Harley and Dr. Phil Krey of the AEC, and a Duquesne Light Co. attorney.

According to Dr. Harley's subsequent memo to the AEC's files [dated June 22] it was a fruitful meeting.

Goldman and DiNunno began by admitting [in a separate memorandum for the files] that someone in N.U.S. had indeed doctored up figures to support the company's position [in past work for the AEC's Health and Safety Laboratory] although there were unfortunately no laboratory records to verify the fact. This aside, however, they had a wonderful new development to report. In the time since President Jones had talked to the AEC, N.U.S. had found some of the original high samples from Beaver Valley.

Now it would be possible to see if that radioactivity had really been there.

This was indeed fortuitous, especially since these samples were by then nearly two years old and the company did not usually retain its samples for more than a year. Evidently they eluded the original search for samples in February.

According to Dr. Goldman, all the company's employees had been instructed to ransack the premises, and the samples had been turned up by two lab technicians in a storage basement where such samples were not usually kept.

Despite the AEC's earlier misgivings about N.U.S.'s credibility, the legitimacy of these newfound samples was accepted without question. Arrangements were immediately made to have them reanalyzed by the AEC, the EPA, an independent private lab, and N.U.S. It was also decided that N.U.S.'s performance in the reanalysis would serve as a test of whether the company had recovered its competence.

So what happened?

The samples were reanalyzed and no more radioactivity! Some of the samples turned out to be as much as twenty times lower than before, but N.U.S. had got it right this time. Their analytical methods were corrected at last. They were saved. Everybody was saved.

The press was notified.

There were a few loose ends.

N.U.S. had to explain why so many of its measurements had been twenty or more times too high in 1971. The company reviewed its laboratory records again and made a new discovery: all through 1971 there had been systematic errors in several of its analytical methods, all tending to produce only erroneously high readings.

That was it. The case was closed.

NUS's safety work for thirty-four other reactors, and even the low readings it somehow managed to obtain at various times and places in Beaver Valley, was allowed to stand unchallenged. Dr. Goldman and DiNunno fired several employees, including the lab chief, who never stopped defending his measurements, and N.U.S. has since continued in its work of making nuclear power plants "absolutely safe to public health."

None of this, of course, was known either to me or the members of the fact-finding commission when the hearings began on July 31, 1973 in the town of Aliquippa. The panel appointed by Governor Milton J. Shapp and chaired by Dr. Leonard Bachman, the Governor's Health Services Director, consisted of seven members in addition to the chairman, representing a broad range of disciplines and wide experience in matters related to public health. Only five of the panel members, however, were independent university-based scientists outside the state government, and only three of these had personal experience with studies of radiation effects in man.

Of the three, Dr. Karl Z. Morgan, Neely Professor of Health Physics at the School of Nuclear Engineering, Georgia Institute of Technology, editor-in-chief of the journal *Health Physics,* first President of the International Radiation Protection Association, and Director of the Health Physics Division of the AEC's Oak Ridge National Laboratory from 1944 to 1973, had the longest association with the problems of radiation, its control, and its measurement.

Next in the length of his professional involvement with radiation and its effects on man was Dr. Edward P. Radford, Professor of Environmental Medicine at the School of Hygiene and Public Health, Johns Hopkins University, who had recently served on the National Academy of Science's Committee on the Biological Effects of Radiation.

The third scientist with recent experience in the evaluation of the effects of radiation on human populations was Dr. Morris DeGroot, Professor of Mathematical Statistics and Chairman of the Department of Statistics at Carnegie-Mellon University.

Of the other two university scientists, one was Dr. Paul Kotin, Provost and Vice-President of the Health Science Center and Professor of Pathology at Temple University in Philadelphia, formerly Director of the National Institute of Environmental Health Science, with a special interest in the environmental causes of cancer, and a consultant to both the National Cancer Institute and the Environmental Protection Agency.

The other member of the scientific panel was Dr. Harry Smith, Jr., Dean of the School of Management at Rensselaer Polytechnic Institute at Troy, New York, who was a biostatistician active in the health field over many years, serving as consultant to the National Center for Health Statistics of the U.S. Department of Health, Education and Welfare.

Also serving on the Governor's Commission was the Secretary of Health for the State of Pennsylvania, J. Finton Speller, M.D., and the Secretary of the Pennsylvania Department of Environmental Resources, Maurice K. Goddard.

Although this was not known to me at the time, it would actually be the staffs of these two state officials who would prepare the final report, since there was no provision for any funding of an independent staff responsible only to the scientist members of the committee. In particular, the radiological portions of the report were to be drafted by Thomas M. Gerusky, Chief, Office of Radiological Health, and Margaret A. Reilly, Chief of Environmental Surveillance in Gerusky's office, both of whom reported to Secretary Goddard. The sections of the report dealing with health effects were to be prepared by Dr. George K. Tokuhata, an epidemiologist recently appointed as Director of Program Evaluation in the Department of Health. All three of these key individuals had in the past made public statements denying the validity of my findings on low-level radiation effects from fallout and releases from nuclear plants. As Griffiths later learned in a series of interviews with some of the commissioners also published in the *Beaver County*

Times, the final report kept being delayed again and again because the staff kept creating drafts which reflected the view that there were no serious problems connected with Shippingport, and which the commissioners were unwilling to sign.

But on the day of the hearings, I was very hopeful that at long last an eminent group of concerned scientists and public health officials would provide the kind of scientific jury able to evaluate fairly the serious evidence for unreported releases and disturbing increases in mortality rates that had recently come to light.

After Dr. Bachman had opened the hearings and introduced the members of the panel, I summarized the data I had previously submitted in two reports to the governor in a series of slides. In addition, I presented further evidence on the changes in mortality rates involving other chronic diseases besides cancer in a number of towns along the Ohio. Thus, in East Liverpool, 5 miles downstream from Shippingport, heart-disease mortality had risen some 100 percent from its low point of 370 per 100,000 deaths in the period 1954–56 to 730 by 1971, while Ohio as a whole had remained constant at about 370 to 390 throughout this period. Yet back in the early 1950s, before Shippingport had started, there was more ordinary pollution from chemicals and coal burning in the Ohio River, from which the drinking water for East Liverpool originated. And in the ensuing two decades, there had been major efforts to clean up the air and water.

I then presented other recent data in support of the possibility that the action of radioactive fallout on all aspects of human health may have been seriously underestimated, thereby explaining the unexpectedly sharp rises in both infant mortality, cancer, and chronic diseases in Aliquippa and nearby river towns since the nuclear plant had gone on line.

Some of this data came from an extensive collection of heath statistics gathered by Dr. M. Segi at the School of Public Health, Tohoku University, Sendai, Japan, from work sponsored by the Japanese Cancer Society. It showed that many types of cancers known to be caused by radiation rose sharply all over Japan, and not just in Hiroshima and Nagasaki, beginning some five to seven years after the bombs were detonated. Thus, while pancreatic cancer had been level for a period of more than ten years prior to 1945—during a period of rapid industrialization, production of chemicals, and growth of electric-power generation by coal—it shot up some 1200 percent by 1965, and only recently began to slow down its enormous rate of climb following the end of major atmospheric bomb testing. The pancreas is also the organ involved in diabetes, a disease that had also shown sharp rises not only in Japan but in the United States, and specifically in the Beaver County area.

Similar patterns emerged from plots of Dr. Segi's data for prostate cancer and lung cancer, the former rising to 900 percent of its pre-1945 incidence, and the latter to 750 percent. And again a similar pattern had taken place near Shippingport, where lung cancer for the nearest sizable town of Midland had risen 500 percent from its 1957–58 rate of 22 to a high of 132 per 100,000 population by 1970, while it had risen only some 70 percent, from 22 to 38 per 100,000 in Pennsylvania as a whole during the same period.

Again, these patterns could not simply be blamed on cigarette smoking alone, although it was known that uranium miners who smoked had some five to ten times the lung cancer mortality rate than those who did not, so that those who both worked in the mines and smoked showed a twenty-five- to hundredfold greater risk of dying of lung cancer as compared with those who neither smoked nor were exposed to the radioactive radon gas. Thus, in effect, the releases of radioactive gases into the already polluted air of Midland has produced the same kind of synergistic effect, as if the people in that town just a mile away from the Shippingport plant had suddenly started to work in the uranium mines.

Thus, the data for the changes in cancer rates in the area for which levels of radioactivity in the air, the water, the milk, and the total diet had been measured as comparable with the levels produced by fallout from bomb tests in Siberia and the Pacific drifting over Japan during the 1950s clearly supported the reality of the data gathered by the N.U.S. scientists recently, and also the reality of the existence of much-higher-than-reported releases from Shippingport in the past.

In further support of the argument that relatively low doses of radiation from nuclear reactor releases can have readily detectable results on human health, I summarized the evidence that infant mortality in Beaver County and other areas along the Ohio had increased in 1960 and 1961 following an accidental release of radioactive isotopes in the course of a fuel-element melt-down at the Waltz Mills nuclear reactor on the Youghiogheny River, some 20 miles upstream from the city of McKeesport in April of 1960.

Within a year after that little-known accident, infant mortality rates doubled in McKeesport and then slowly declined again to the level of the rest of Allegheny County, which gets its drinking water mainly from the Allegheny River. And the effects could be seen in a steadily declining pattern of infant mortality peaks along the Monongahela and Ohio River communities for 160 miles downstream.

In the course of the questioning period that followed my presentation, I was asked how it was possible that such relatively small doses comparable to normal background levels could lead to such large changes in mortality rates, when it apparently took ten to a hundred times these levels to double the risk for the survivors of Hiroshima and Nagasaki. In response I cited the startling results of a recent study published in the journal *Health Physics* in March of 1972 by a scientist working for the Canadian Atomic Energy Laboratories in Pinawa, Manitoba, Dr. Abram Petkau. Dr. Petkau had been examining the basic processes whereby chemicals diffuse through cell membranes. In the course of these studies, he had occasion to expose the membranes surrounded by water to a powerful X-ray machine, and observed that they would usually break after absorbing the relatively large dose of 3500 rads, the equivalent of some 35,000 years of normal background radiation.

This certainly seemed to be very reassuring with regard to any possible danger to vital portions of cells as a result of the much smaller doses in the environment from either natural or man-made sources. But then Dr. Petkau did something that no one else had tried before. He added a small amount of radioactive sodium salt to the water, such as occurs from fallout or reactor releases to a river, and measured the total absorbed dose before the membrane broke due to the low-level protracted radiation.

To his amazement, he found that instead of requiring a dose of 3500 rads, the membrane ruptured at an absorbed dose of three-quarters of one rad, or at a dose some 5000 times less than one rad, much less than was necessary to break it in a short, high-intensity burst of radiation such as had occurred at Hiroshima and Nagasaki.

Dr. Petkau repeated this experiment many times in order to be certain of this disturbing finding, and each time the result confirmed the initial discovery: the more protracted the radiation exposure was, the less total dose it took to break the membranes, completely contrary to the usual case of genetic damage, where it made no difference whether the radiation was given in one second, one day, one month, or one year.

By a further series of experiments, he finally began to understand what was taking place. Apparently a biological mechanism was involved in the case of membrane damage that was completely different from the usual direct hit of a particle on the DNA molecules in the center of the cell. It turned out that instead, a highly toxic, unstable form of ordinary oxygen normally found in cell fluids was created by the irradiation process, and that this so-called "free radical" was attracted to the cell membrane, where it initiated a chain reaction that gradually oxidized and thus weakened the molecules composing the membrane. And the lower the number of such "free radicals" present in the cell fluid at any given moment, the more efficient was the whole destructive process.

Thus, almost overnight, the entire foundation of all existing assumptions as to the likely action of very low, protracted exposures as compared to short exposures at Hiroshima or even from brief, low-level medical X-rays had been shaken. Instead of a protracted or more gentle exposure being less harmful than a short flash, it turned out that there were some conditions under which it could be the other way around: The low-level, low-rate exposure was more harmful to biological cells containing oxygen than the same exposure given at a high rate or in a very brief moment.

No longer was it the case that one could confidently calculate what would happen at very low, protracted environmental
exposures from studies on cells or animals carried out at high
doses given in a relatively short time. It was clear that the
direct, linear relation between radiation dose and effect was
no longer the most conservative assumption, for it was based
on the implicit assumption that a given dose would always
result in a given increase in risk, no matter whether the radiation
was absorbed in one second or one year. Clearly, if Dr. Petkau's
findings were to be confirmed by other experiments in the future, our whole present understanding of low-dose radiation
effects would have to be revised, since small exposures might
turn out to be far more harmful to living cells than we had
ever realized.

Thus, I pleaded we should not reject evidence for much
higher than expected infant and cancer mortality rates merely
because that evidence did not seem to agree with our previous
estimates based on high-level, high-rate exposures at Hiroshima
and in various studies. I now believed that we had to be prepared
to revise drastically our expectations as to what apparently
innocuous low-level, chronic radiation exposures to critical cells
and organs from environmental sources might do.

My own testimony was followed by that of Dr. Irving Bross,
a well-known biostatistician from the Roswell Park Memorial
Cancer Institute in Buffalo, New York, who had himself been
studying the effect of low-level radiation on childhood leukemia
for many years. In summarizing his findings Dr. Bross stated
that there exists a wide range of individuals with very different
degrees of sensitivity to radiation, depending upon their age
and their past medical history.

This fact alone would invalidate any estimate of the likely
effect of small radiation exposures to a large human population,
since these had been based on the average adult, obtained at
high doses, and on the assumption of a linear relationship between dose and effect. For a non-homogeneous group, the more
resistant individuals such as healthy young adults would not
show any significant effects, while either the very young or
the very old and those with immune deficiencies, allergies, and

other special conditions might show an unexpectedly large effect. As Bross had put it in a letter to *The New York Times* published just a few weeks before he testified: "It follows that procedures for calculating 'safe levels' based on 'average exposures' of 'average individuals' are not going to protect the children or adults who need the protection most."

Next was the testimony of the Deputy Director of the Division of Biology and Medicine of the U.S. Atomic Energy Commission in charge of all biomedical and environmental research, Dr. W. W. Burr, Jr. This witness, as recorded by the reporter for the *Beaver County Times,* Bob Grotevant, "tabbed all allegations about a definite correlation between radioactive emissions from the Shippingport plant and increased infant deaths and cancer cases made by Dr. Sternglass as 'unsupportable.' " Burr then announced that a number of follow-up tests after publication in 1971 of "erroneous" test data by the N.U.S. Corporation "proved that no such high levels of any radioactive products existed near the plant."

This, then, was the way that had been chosen by the AEC to deal with what had happened, as we were to learn later from the internal memoranda, and one witness after the other for N.U.S., for the utility, for the EPA, and for the Commonwealth of Pennsylvania followed the line agreed upon in the correspondence and secret meetings described in the memoranda. Each independent set of data was rejected as unreliable or meaningless when it showed the existence of high radiation levels or increases in mortality rates.

As Anna Mayo, who covered the proceedings for *The Village Voice,* put it in an article published a few months later, "it was all redolent of—you guessed it—Watergate. In the audience, environmentalists gnashed their teeth, wishing that the Shippingport horrors could have been exposed on national television. If Duquesne Light would cover up, would not Con Ed, LILCO, or Commonwealth Edison do the same if Indian Point, Shoreham, or Dresden were at stake?"

Indeed a great deal was at stake: In 1973 some thirty-eight new nuclear reactors were in the process of being ordered, the largest number ever in one year, each representing a potential business of about a billion dollars. And it was the stated aim of the Nixon administration and the nuclear industry to see a thousand of these reactors operating near the cities of our nation by the end of the century. It would indeed be difficult for any human beings not to have minimized the danger when a thousand billion dollars were at stake.

As expected, when the report of the Governor's Commission finally appeared a year later, after the licenses had been granted to Beaver Valley Unit I and II, it did not call for a moratorium on nuclear power plants, as Anna Mayo had suggested it should at the end of her article. In fact, she had predicted the outcome exactly. As she had put it bitterly: "About the most that can be expected is a modest plea for further studies: that is, more and more necrophiliac nitpicking."

The summary of the commission's report set the tone of the entire document. By carefully using certain qualifying words that are easily passed over by the hurried reader, such as "substantial," "systematic," or "significant," a draft had finally been prepared by Tokuhata, Gerusky, and Reilly that the members of the committee could no longer continue to refuse to sign after months of efforts to arrive at some sort of acceptable wording. It provided sentences which, when taken separately, could be widely used by the utility to claim that it had been completely cleared. For example, consider the very first sentence: "There is no substantial evidence that the quantities of radioactive materials released by Shippingport Atomic Power Station have been greater than reported by the plant operators." This sentence was followed, however, by one that would satisfy the consciences of some of the more concerned commissioners: "However, the absence of comprehensive off-site monitoring during plant operations precludes accurate verification of the data on plant releases," and so on throughout the long and inconclusive report.

Far more revealing than the report as to the true feelings of four of the five independent scientists on the commission willing to go on record were the answers to questions submitted to them by Griffiths in his article, which appeared just before Governor Shapp released the report in June of 1974.

For instance, to the question, "Did the data in the original N.U.S. report point to Shippingport as the source of the high radiation data," the scientists answered as follows:

DR. DEGROOT: "If we accept those data, then the circumstantial evidence points to Shippingport largely because of the location of the radioactivity and the lack of plausible alternate sources."

DR. MORGAN: "The original N.U.S. data very strongly suggested to me that the radioactivity came from the plant. If you take the data as fact, you'd be very hard-pressed to find any other source that could explain it."

DR. RADFORD: "Well, there was some indication in the original N.U.S. data that there was a release from some source. As to whether that source was Shippingport, I'd have to look up the data again."

DR. SMITH: "I can't find any direct connection between the radiation levels measured by N.U.S. and the Shippingport plant. All that mish-mash is so unscientific that one would never be able to draw any valid scientific inferences from it."

Another question referred to the discrepancy between the original N.U.S. analysis and the reanalysis: "After N.U.S. reanalyzed its data, the high radiation levels disappeared. Did this reanalysis prove to you that the radioactivity was never there?"

DR. DEGROOT: "No, it did not. It did convince me that the reanalysis was highly unreliable. However, I am equally convinced that the original N.U.S. data showing high levels cannot be considered reliable evidence. There are just so many inconsistencies in their work that I cannot accept any of it. . . . This comment does not mean that all their high readings were wrong. In fact, I find it highly unlikely that N.U.S. could have made systematic errors, all in one direction, in several different analytical techniques."

DR. MORGAN: "The explanations advanced by N.U.S. did

not at all convince me. For example, if they had found something wrong in only one of their systems, it would not be too surprising. We all make mistakes. But to have systematic errors in several different analytical techniques, all tending to produce only high readings— the chances of that are quite low. . . . There appears to be a strong suggestion of dishonesty, and that estimate is borne out by written comments from Dr. John Harley of the AEC, whose integrity I respect. Dr. Harley found that N.U.S. seems to have doctored some of their data to fit their arguments. If a person will do that with one set of scientific data, it is very possible he will do it with another. . . . So, as far as I can see, there is no proof that the radioactivity levels around Shippingport were not quite high in the past. For a long period now the radioactivity levels in milk in that general area have been high according to the public-health agency surveys, which are completely separate from the N.U.S. survey. This has never been explained."

DR. RADFORD: "Well, they had three separate laboratories reanalyze some of the original 1971 milk and soil samples, and each lab got similar low readings. If these samples were valid, then it is pretty clear there was not much radioactivity there to begin with. Now of course you could say they dug up soil from somewhere and analyzed it—I cannot argue that."

DR. SMITH: "I think that the degree of scientific merit on one side really was better. I would accept the explanations advanced by N.U.S."

Another question: "Was there any evidence in the mortality statistics that Shippingport had caused health damage, or did the statistics tend to refute this?"

DR. DEGROOT: "We cannot really decide the issue because of the poor quality of the available health statistics and because the population is not large enough for a really meaningful statistical analysis. But there is certainly nothing in the available data to lower the probability that there may have been health damage. It is true that the Pennsylvania State Health Department went back and discovered errors of a certain type in its published infant mortality rate for Aliquippa in 1971, and that the ensuing corrections sharply lowered the rate. . . . However, I feel it is likely there were also errors of another type which could have raised the rate back up again. Unfortunately, the resources were not available to investigate this possibility. So, to my mind, the corrections are incomplete. The only type of error investigated was one that would reduce the number of deaths and lower the rate. . . . In any case, I think there remain some anomalies that have not been fully explained. For example, I did an analysis of infant mortality in Aliquippa, and the rate definitely seems to have shifted upward recently. To my mind this upward shift is not fully explained by demographic or socioeconomic factors. I do not know if any of it is due to Shippingport, but I think it warrants further investigation."

DR. MORGAN: "I do not personally feel that the mortality statistics refute the possibility of some adverse effects on the population's health. Taking the original published data, it appears to me that there was an effect. However, after the Health Department got through making corrections and applying all the epidemiological and statistical techniques to the mortality rates for the population near the reactor, they seem to have come up with the belief that there were no significant health effects. . . . I cannot help but be a little skeptical. To me, if you are going to make all these corrections for the population that might have been exposed to radiation, you have to give equal consideration to the unexposed control population. It was very obvious to me that if they had, it would have made a difference in at least one instance."

DR. RADFORD: "The statistical evidence favors the hypothesis that the plant did not cause any health damage. For example, the mortality rates do not decline with distance in all directions away from the plant. The mortality rates for Beaver County as a whole are quite low, and on that basis one would be hard-pressed to say that Aliquippa was affected, since the rest of the county should also be high. . . . Then, when the mortality rates for Aliquippa are corrected for errors, you see that Aliquippa is no worse off than any other town with comparable population characteristics."

DR. SMITH: "In my opinion the mortality statistics indicate there was no effect from the reactor. The adjusted mortality rates are not abnormally high. One comes to the conclusion that the Shippingport area may not be the greatest place to live, since the mortality rates are higher there than in many other communities, but such high rates are normal, expected occurrences in places with the kind of demographic and socio-economic characteristics you find around Shippingport. . . . Also, I have to find a scientific link between radiation exposure and infant mortality, and this requires a great deal of what I call logical extrapolation or inferences step by step through a process which proceeds from the birth of a child to its ultimate death, and I cannot find sufficient evidence for that link in this case."

Although the majority clearly were deeply suspicious of the "reanalysis" of the radiation data and the "adjustment" of the vital statistics by Tokuhata, I was surprised by Radford's comment that the mortality rates do not decline with distance away from Shippingport, and that therefore the evidence favored the hypothesis that the plant did not cause any health damage.

Not until later, when I saw the final report, did I see what could have led Radford to this conclusion. In Table 13, Tokuhata had listed the cancer death rates according to distance from Shippingport for the years 1961 to 1971. There were columns for the rates within 5 miles, between 5 and 10 miles, beyond 10 miles, for Beaver County, and for Pennsylvania as a whole. And at the bottom of each column, there were listed the average mortality rates for each of these regions.

When I looked at them, I was startled to find that Radford seemed to be right. The lowest rate did in fact exist for the circle 5 miles in radius around Shippingport: 155.7 compared with 170.4 in the next, more distant region 5 to 10 miles away from the plant, and a still higher rate of 182.3 for Pennsylvania as a whole. This certainly seemed to suggest that radiation was good for one's health, and that the closer one lived to the reactor, the better off one would be.

What exactly had Tokuhata done to arrive at this conclusion that had obviously convinced Radford and Smith? It took me a while to work it out, but when I did I was furious. Looking down the entries for each year from 1961 to 1971, I saw that all areas showed lower cancer rates in 1961 than in 1971, but that the area nearest to Shippingport had happened to have by far the lowest rates to begin with, well before any major releases had occurred from Shippingport and well before any increases in cancer mortality due to Shippingport could have shown up in the statistics. It had been a largely rural area, relatively free from pollution and therefore with relatively good health, cancer mortality having reached a low point of only 102.6 per 100,000 population in 1964, lower than any other listed at any time for any area in the table. The average for the first four years, 1961–64 was only 133.4, compared with 155.3 for the 5-to-10 mile range and 176.8 for Pennsylvania as a whole.

But by the time that the 1963–64 Shippingport releases had had a chance to act, namely by 1969–70, the area nearest to Shippingport had increased the most, shooting up to a peak of more than double its lowest rate of 102.6, namely to 225.6 in 1969 and 218.9 in 1970, while the more distant areas increased much less. Thus, the 5-to-10-mile-distant zone had risen to 189.2 by 1969 and 191.2 by 1970, while the area of Beaver County beyond 10 miles from Shippingport was listed at only 164.9 and 164.3 for these years.

In fact, taking the last four years of 1968 to 1971 in the table when cancers had had a chance to manifest themselves, and comparing them with the first four years when the effect of any releases could not yet have appeared in the mortality statistics, it was clear that the data fully confirmed my earlier findings obtained from the Vital Statistics reports of Pennsylvania and Ohio by town and by county. The greatest increases had indeed taken place for the people nearest to the plant: a rise of 38 percent compared with only 22 percent for the next zone and 20 percent for the area beyond 10 miles, while Pennsylvania as a whole showed only a 6 percent increase in cancer mortality.

Thus, by averaging over all the eleven years listed in the table so as to include the years of lowest cancer rates for the rural area around Shippingport before the plant could have had any effect on cancer rates, Tokuhata had successfully managed to give the impression that the closer one lived to the plant, the less was the risk of cancer.

There was one question that had remained unanswered even by the internal documents from the AEC files: How and where in the plant did the radioactive gases escape without being officially reported, as required by both state and federal regulations?

As so often before in the Shippingport story, the answer came in the most unexpected manner, this time not through the mail but in a phone call late one evening a few weeks after the Aliquippa hearings had ended.

The caller said that what had been brought out at the hearings so far was in the right direction, but that the full story behind the high radioactivity in the area could be found by putting the plant operators on the stand in the forthcoming licensing hearings that were to be held by the AEC later in the year. What we needed to do was to have the men explain during cross-examination the details of the treatment system for the radioactive gases, and then force them under oath to say whether they had found any anomalous conditions in the hold-up tanks where the radioactive gases were supposed to be stored for many weeks to allow the shorter-lived radioactivity to decay before they would be discharged from the monitored stack.

This was of course the kind of break we had hoped for.

Together with the internal memoranda of the AEC that had
revealed the attempt to explain away the findings of high radioactivity in the air, the soil, the milk, the water, and the local
diet, it would complete our case for arguing that the Duquesne
Light Company should not be given a license to operate two
even larger nuclear reactors, since their employees were either
too incompetent or too corrupt to do so without endangering
the health and safety of the public.

And so I obtained the detailed engineering drawings of
the gas-treatment system for the Shippingport plant from articles published in the literature, and explained the complex system to the attorney for the city, Al Brandon, who would have
to do the actual cross-examination.

The hearings by the Atomic Safety and Licensing Board
on the operating permit for Beaver Valley Unit I and the construction permit for Unit II finally got under way in the fall
of 1973 in the Federal Court House in Pittsburgh. Although
we had few illusions as to what the ultimate decision would
eventually turn out to be, we at least hoped to expose to the
public what had actually been going on behind the scenes at
the Shippingport plant, widely advertised all over the world
by Westinghouse and Duquesne Light as the cleanest and safest
nuclear reactor in the world.

For a while we did not know whether we would be allowed
to put the operators of the plant on the stand. But then the
ruling came down, and it all really happened.

The first few men, when shown the diagrams of the gas-
treatment system, claimed that they were not aware of anything
abnormal. But suddenly, one of the men, when pressed by
Brandon as to whether he had ever noticed anything unusual
in the operation of the system, and whether there might not
have been some leakages from the gas-storage tanks in the
yard, admitted that he had observed something that had caused
him to become concerned.

Some time in late 1970 or early 1971 he had noticed an
unusual drop in the amount of recorded radioactive gas releases
in the plant log, and he had mentioned it to his supervisor,
who told him not to worry about it. Questioned by Brandon

he admitted that the situation persisted over a period of a few weeks, and that he then decided to investigate what might be going on for himself. He went out into the yard where the large gas-storage tanks were located and found that a lock on one of rusty valves had been broken. The valve looked as if it might be leaking. Using a small brush to paint a soap solution over the suspected area, he saw bubbles being formed, indicating that radioactive gas was in fact leaking from the tank.

Again, he said that he reported the situation to his supervisor, who told him that he would take care of it, and that he should not concern himself with this problem any more since this was not part of his job.

As Brandon expected, none of the supervisors he put on the stand could recall this incident, and the local newspaper that evening reported that the plant personnel had testified that there were no problems in the plant.

Dr. Morton Goldman, the vice-president of N.U.S. and former public-health officer in the U.S. Department of Health, Education and Welfare, testified under oath that all their early high readings of radioactivity had been in error, substantiating the testimony of the plant supervisors that no unusual or unreported releases could have taken place, and a few months later the Atomic Safety and Licensing Board issued the permits for the new reactors.

Once again, the industry had managed to win the battle in the special courts set up by the AEC, which controlled the judges, the staff, and the rules of procedure for the benefit of the industry it was designed to promote and protect.

It was only the people that were the losers. Two years after the licenses were granted and five years after the high radiation levels had been measured by the N.U.S. Corporation, with the same time delay as in Hiroshima, the cancer rates in Beaver County and Pittsburgh climbed to a second peak. They rose a full 23 percent in Beaver County and an unprecedented 9 percent in Pittsburgh in the course of only three years: The rise to an all-time high of 304.8 per 100,000 population took place after a generation of costly efforts to reduce the ordinary pollution from fossil fuels in the air and chemicals in the water.

But the heaviest price of all was to be paid by the men who worked at Shippingport, as I was to learn at another kind of hearing at Aliquippa seven years later.

When preparing testimony for a hearing before a workmen's compensation referee in behalf of the family of a man who had died of bone-marrow-type leukemia while working at the Beaver Valley nuclear plant next to the old Shippingport reactor, I was shown the death certificates of twenty-one other operating engineers who had died between 1970 and 1979. All of them had been working with pumps and other heavy equipment to clean up the radioactive spills and move the radioactive wastes on the site. Out of these twenty-two men, ten had died of cancer, more than twice the number normally expected.

Even more significantly, four of these ten were of the bone-marrow-related type, namely multiple myeloma and myelogenous leukemia, known to be most readily induced by radiation, when less than one in twenty cancers of this type would have been expected.

The men who worked at Shippingport were only too well acquainted with these facts. There was a common saying among them: high pay and early death.

Yet there was also a sign of hope for the future. After Shippingport was shut down by an explosion of hydrogen gas in its electrical generator early in 1974, infant mortality in the town of Aliquippa declined to an all-time low of only 11.3 deaths per thousand babies born in 1976.

If the public could only learn these facts as the nation entered the third century of its revolution against the arbitrary authority of another distant government careless of the inalienable human rights to life and liberty, even the tragic tide of rising cancer and damage to the unborn could eventually be reversed.

16

The Minds
of the Children

DRAINED by the long battle to warn the people of Beaver County and Pittsburgh of the dangers arising from the "normal" operations of nuclear plants, I decided to devote myself again to my much-neglected research in physics and radiological instrumentation. Many people had by now taken up the fight to warn the public about other previously unrecognized dangers of the nuclear fuel cycle, all the way from the mining and milling of uranium ore to the ultimate disposal of the long-lived radioactive wastes and the possible theft of plutonium by terrorist groups to make home-made bombs. Thus, when Henry Kendall of the Union of Concerned Scientists joined forces with Ralph Nader in exposing the previously hidden risks of a major accident in early 1974, I felt that the battle was in good hands while I caught up with my other responsibilities.

Now and then I would of course be asked to speak or testify at licensing hearings or court cases, but with the revelations of secret abuses of power in the Nixon administration and the formation of Nader's "Critical Mass" organization of anti-nuclear groups in the fall of 1974, I had the feeling that the public's blind trust in its government leaders had at last been shaken, and that the tide had begun to turn. It seemed to me that the nuclear juggernaut would eventually be halted as a new generation of young scientists, engineers, and political leaders born after Hiroshima could take an unbiased look at the enormous problems that had been kept from the public and that were now increasingly coming to light.

What did continue to concern me very much were the repeated episodes of heavy fallout of radioactive iodine from Chinese nuclear tests that continued to damage the thyroids of the unborn and the infants, but that continued to be downplayed by state and federal health agencies. There was also the failure of any real progress toward an end to the multiplication of nuclear bombs, despite the signing of the SALT treaty that at least for the moment had halted an all-out anti-ballistic missile race.

In addition, there were also the continuing underground bomb tests by the U.S., the Russians, the French, and the British. No one paid much attention to these anymore, but on numerous occasions in the past such tests had spewed forth radioactive gases that kept raising the risk of cancer and threatening the life and health of the newborn thousands of miles away.

I remembered only too well the tragic story of one of the worst of such accidents. On December 18, 1970, more than seventeen years after Professor Clark and his students accidentally discovered the rainout in Troy, New York, the U.S. Atomic Energy Commission conducted an undergound nuclear-weapons test at the Nevada Test Site. Code-named "Baneberry," this Hiroshima-sized bomb was exploded some 800 feet underground. The explosion opened up a fissure in the rock, and large quantities of radioactivity escaped upward into the atmosphere. In the vicinity of the test site, the slowly drifting cloud of radioactive dust produced readings of 25 rads per hour on the ground. Hundreds of employees were seriously overexposed and had to be quickly evacuated.

Shortly after the accident, the AEC's Division of Biology and Medicine and the Utah State Division of Health notified a team of fallout specialists working at the University of Utah under contract to the AEC. The team, headed by Drs. Robert C. Pendleton and Charles W. Mays, scientists who had long warned of the dangers to the infant thyroid from radioactive iodine, immediately set about determining the direction and intensity of the radioactive clouds. From the Salt Lake Weather Bureau they learned the speed of the winds at the time of detonation, while the Nevada Operations Office informed them that the radioactive leakage had occurred during a period of very strong wind shear, with winds at different altitudes blowing in different directions. The team was able to estimate that during the next twenty-four hours the lowest part of the cloud would probably go to the east of Winnemucca, Nevada,while the layer above would be blown across Utah to the southeast. The next-highest layer was apparently headed for New Mexico, while the topmost parts were expected to be carried into Utah between Highways 56 and 21.

Once the direction of the fallout was estimated, the extensive network of sampling stations around the state, constructed in the years since the dangers from fallout in the milk were discovered, could go into operation, estimating the strength of the radioactivity in the air and on the ground. This was indeed a far more sophisticated operation than the one mounted so long before by Dr. Clark's students, driving from town to town in their jalopies carrying rudimentary Geiger counters. And in this case, there was no problem about estimating the internal radiation dose, something that had not even been considered in Troy back in 1953. Now, a program run jointly with the Utah Division of Fish and Game was put into action to procure samples of the local wildlife that had been in the path of the fallout. Conservation officers were enlisted to collect deer that had recently been killed on the highways, while sheep, deer, and rabbits were shot in areas where they were regarded as a nuisance. These animals were to be dissected in the laboratory, where the concentrations of isotopes in their body organs would be measured.

On December 19 and 20, teams departed from Salt Lake City to obtain samples of snow and vegetation. Instead of the burdock leaves favored by Dr. Clark and his students, the fallout specialists took samples of alfalfa, sagebrush, and juniper. Some of the alfalfa samples consisted of loose hay from exposed bales.

According to the laboratory results, the most prominent gamma-radiation-emitting isotopes in the Baneberry fallout were the short-lived, intensely radioactive iodine 131 and iodine 133. These were found in the lungs, thyroids, stomachs, and fetuses of deer and sheep, as well as in snow, milk, and vegetation. The fallout specialists were able to determine that if the Baneberry explosion had happened to take place in the warmer months, when the cows were out to pasture, the total dose to the local children from iodine in the milk would have been approximately 120 millirads, with some receiving much higher exposures. This did not include the dose from the other isotopes inhaled or eaten. Furthermore, the scientists observed that low wind velocities and an atmospheric inversion had fortunately served to keep the fallout that reached Utah fairly stationary for a number of days in a position over relatively unpopulated areas. This allowed quantities of the heaviest fallout particles to settle to the ground, while the short-lived isotopes lost much of their radioactivity before the cloud was blown over the more populated areas. If these fortuitous circumstances had not existed, the doses might have been much higher for the people of Utah, perhaps as high as in Troy in 1953.

As it turned out, the predictions of where the fallout would drift were wrong. The heaviest clouds went north and northeast toward Idaho, Washington, and Montana, where rain and snow brought down much more radioactivity than in Utah. Subsequently, the fallout from Baneberry was detected across the northern U.S. by large rises in the cesium 137 levels in milk, as could be seen in the state-by-state tabulation of cesium levels for December 1970 printed in the April 1971 issue of *Radiation Health Data and Reports*. Data for the radioactivity on the ground also showed that the fallout had drifted into Canada, thus violating the provisions of the 1963 test-ban treaty, which does not permit nuclear tests that release radioactivity beyond the national borders of the nation conducting them.

But as far as the general public was concerned, there was only the following statement by the AEC, carried by *The New York Times* and the rest of the press around the country:

The Commission said that the radioactivity was of such low intensity that it presented no danger. It was detected at altitudes of several thousand feet, and only the most minute traces of radioactive contamination would reach the ground, the Commission said. The AEC . . . said that no-where outside the immediate area of the test was the fallout dangerous to human life or health.

In the spring of 1971 our group gathered the data for radio-activity in the air, in the milk, and on the ground both before and after the Baneberry test. This was then correlated with the mortality figures for infants born following the explosion, as reported in the U.S. Monthly Vital Statistics. In all of the states where the total radioactivity rose highest—Idaho, Montana, Oregon, Nevada, Washington, Nebraska, and as far away as Minnesota and Maine—infant mortality also rose sharply during the first three months after the test. Across the rest of the U.S., the pattern of general decline continued.

It was shortly after reading another story in the papers about how the United States and the Soviet Union had failed to agree once again on a treaty to halt all underground nuclear tests that my attention was caught by an article in *The New York Times* about an apparently unrelated subject. The report dealt with the fact that in 1975 the scores in the nationwide Scholastic Aptitude Tests had dropped by the largest amount in two decades. While there had been a more or less steady decline in both the verbal and the mathematical scores since the mid-1960s, generally by no more than 2 or 3 points, the average verbal scores had suddenly dropped 10 points in a single year. Since our son was taking the S.A.T. tests that year, I read the article with more than casual interest.

Suddenly the question flashed through my mind: When were these young people born or in their mother's womb? Most of them were 18 years old when they graduated from high school. What was 18 taken from 1975? It was 1957, the year when the largest amount of radioactive fallout ever measured descended on the United States from the highest kilotonnage of nuclear weapons ever detonated in Nevada. Just as in the case of the Baneberry test, the radioactive iodines must have gone to the thyroids of the infants in their mother's womb, where it would retard their growth and development ever so slightly so that it was not readily noticeable, and only when the children were tested 17 to 18 years later on a nationwide scale would it show up in a sharp drop in intellectual performance.

Clearly, if the effects were serious enough to lead to a rise in infant mortality and congenital defects back in 1957, as I knew had taken place, then for every baby that died shortly after birth, there must have been many who were minimally brain-damaged or whose cognitive growth may not have reached its full potential.

I remembered from the 1969 Hanford symposium that this was exactly what had happened to the young children on the Marshall Islands after the radioactive cloud from the "Bravo" hydrogen-bomb test in 1954 had accidentally showered the island of Rongelap, 150 miles away, with fresh fallout. As reported by Conard at that meeting, in the following fifteen years, all the children developed thyroid disease of one form or another and showed severe growth retardation, both in their bodies and the size of their brains.

But the thought was really too disturbing to contemplate in all its enormous implications. Perhaps it was just a coincidence and nothing more. After all, as the *Times* story made clear, there were so many other possible factors that could have been involved, including a deterioration of the schools, more disadvantaged students taking the tests, more urban problems, and the whole upheaval of the Vietnam war. Even too much television viewing had been blamed for the drop in reading ability, as well as a general decline in motivation among young people. But I was glad that I had urged my wife and all our friends to give powered milk to their children during their years of infancy, in which the short-lived iodine 131 had had a chance to decay away.

Not being an expert in the field of psychological testing,

I clearly was out of my depth. Consequently, I decided to put the idea aside for the moment, thinking that perhaps some day there might be an opportunity to discuss it with colleagues and friends more knowledgeable in this field. Besides, I saw no obvious way to test this idea further at this time. It was true that the decline had begun only in 1963, eighteen years after 1945, when the first bombs were detonated and infant mortality began to halt its decline. But only time would tell whether the decline would end when the students taking the test were those who were born during the temporary nuclear bomb test moratorium between 1959 and 1961. (This group would be taking the S.A.T. in 1977 and 1978.)

In the face of widespread public alarm, a special panel on the decline in Scholastic Aptitude scores was created under the direction of former Secretary of Labor Willard Wirtz. The Wirtz panel commissioned more than two dozen special research studies under the joint sponsorship of the College Board and the Educational Testing Service; these studies were published in 1977 together with a summary report, a copy of which I sent for when its completion was announced in another article in *The New York Times*.

It was clear from the report that despite this major effort to identify the cause or causes of the disturbing decline in test scores, no single factor or group of factors seemed to explain the observed pattern of decline. The various studies did conclude, however, that certain factors were *not* likely to have played a significant part in the sudden decline. For example, cultural bias, differences in the predictive ability of the tests for whites and blacks, changes in the difficulty of the test, and tests getting out of line with high school practices and standards were eliminated as likely explanations.

Since even after this major research effort no one had come up with any really adequate explanation, I decided to discuss my hypothesis with an old friend, Dr. Henry P. David, a psychologist living near Washington, whenever the opportunity would present itself next.

To my surprise, after listening quietly to my explanation for a long while, he did not think that it was so impossible after all. There had apparently been a growing recognition in the psychological community that many physical factors acting on the developing baby during pregnancy—including cigarettes, alcohol, drugs, and anesthetics used during delivery—could result in retarded growth, underweight births, and various degrees of learning and behavioral problems later in childhood.

Thus encouraged, I decided to pursue the matter further, particularly since one of the predictions of the hypothesis had just been confirmed: As suddenly as the annual drop in verbal S.A.T. scores had gone to ten points in 1975, it had just as unexpectedly stopped dropping so precipitously in the following two years, declining by only three points in 1976 and a mere two points by 1977.

Many years ago, I had had occasion to look up the total amount of fission energy released by small tactical weapons detonated in Nevada as measured in kilotons, or thousands of tons, of equivalent weight of TNT. (The Hiroshima bomb was approximately 15 kilotons.) When I found the figures, they were 303 kilotons in 1957, 18 kilotons in 1958, and none in 1959. So far, at least, the idea had withstood its first test, and the ability of a theory to make a correct prediction is universally regarded as a very crucial factor in accepting it. But I needed to find someone in the field of educational testing with whom to work on the further examination of these ideas.

It so happened that shortly after I had started to look for a collaborator, I received a letter from Dr. Steven Bell, an educational psychologist at a small college in Georgia. In the letter he told me that he had heard about my findings on infant and fetal mortality changes that were correlated with fallout, and he asked whether I had ever considered the possibility that it might have had an effect on learning ability.

Delighted with this coincidence, I wrote back that indeed I had begun to suspect this, and that I would be happy to work with him on this question. I included copies of all my relevant papers and some preliminary plots showing a correlation of the declining S.A.T. scores with the pattern of accumulated external gamma-radiation doses measured at the Brookhaven National Laboratories. Oddly enough, this had been introduced into the record of the licensing hearings for the Shoreham nuclear plant some years ago by proponents for its construction when I was asked to testify as to the potential danger that the planned releases might present for children born in the area.

As it turned out, Bell had for some years become increasingly concerned about the possibility that physical and chemical agents in the environment might have much more serious effects on I.Q. and achievement test scores than had been suspected. Such factors acting on the baby in the mother's womb would be indistinguishable from hereditary problems, so that individuals born in an environment where there were such deleterious factors present might show learning disabilities that would wrongly be blamed either on "bad genes" or the poor social and educational environment alone.

Since very often the poor were blacks, Indians, or Puerto Ricans, the generally lower I.Q. and S.A.T. scores of these groups as a whole could mistakenly be blamed on genetic factors when actually there could be an unrecognized effect of such physical and chemical agents as DDT, cigarette smoking, air pollution, herbicides, poor diet, high fallout, and drug use during pregnancy—factors that had been generally ignored by psychologists studying intellectual development in the past. And since poor diet and health are more common among the poor, minority groups would be disproportionately affected by these factors.

Certainly all the earlier studies on infant and fetal mortality rates had consistently shown that nonwhites suffered about twice the mortality rates of infants within the white population, and this was widely recognized as being connected with poverty and poor diet and not with any inherited factors. When the mortality rates had stopped declining after the onset of heavy nuclear testing in the early 1950s, the greatest negative effects were in the nonwhite population of the rural south and the urban ghettos of the Northeast, where not only the fallout but also the poverty was greatest. And even after the end of the heaviest fallout from the atmospheric bomb tests in the mid-1960s, although both white and nonwhite infant mortality were declining rapidly once more, as I had predicted back in 1969, the absolute mortality rate for nonwhites was still twice what it was for the white populations in such urbanized states as Pennsylvania, New York, and Illinois.

But wherever there were no large groups of very poor populations and the environment was relatively free from both ordinary industrial and radioactive pollution—as in Hawaii, Alaska, Montana, and New Hampshire—infant mortality rates were plunging to unprecedented low levels in the late 1970s. The rates were dropping far below those of the urban states with large nuclear reactors, such as New York and Pennsylvania, despite the higher number of physicians per person and the greater access to hospitals with special units for the care of premature babies. And a similar pattern had begun to show in cancer rates.

All of these considerations supported the hypothesis that radioactive fallout might have been one widely distributed factor that had been neglected in the search for a cause of the declining learning and reading abilities of the young students born in the 1950s. But how could this be tested further?

Bell and I had sent for one of the studies carried out for the Wirtz commission by Dr. Rex Jackson of the Educational Testing Service. The study contained a detailed statistical breakdown of the scores by region since 1971. Since fallout also had differed in intensity for different regions of the United States, here was another chance to check the hypothesis.

If it was indeed the poor schools of the large urban areas that were mainly responsible for the drop in scores, then clearly the greatest declines should have taken place in the Middle Atlantic Region, which included New York, Pennsylvania, New Jersey, Delaware, Maryland, and the District of Columbia, together with the Midwest populations in Illinois, Michigan, and Ohio, with their large ghetto areas.

On the other hand the Western Region, which included the states of Alaska, Hawaii, California, Arizona, Oregon, Washington, Montana, Utah, Wyoming, Idaho, and Colorado, with their relatively rural populations and relatively fewer urban ghetto problems should clearly show the least decline.

But exactly the opposite was the case. Comparing the scores

for the high school graduating classes of 1976 with those for 1974, during which the United States as a whole dropped 13 points in the verbal test, the West dropped 19 points compared to only 9 points for the Midwest and 14 points for the Middle Atlantic states.

Even the South, with its relatively low average income, poorer school systems, and lowest educational expenditures, dropped by much less than the West, namely by only 13 points. No other region declined as much as the western United States, with its relatively clean air, clean water, and generally higher average socio-economic level.

Nor would the wide difference in the drop fit the hypothesis that it was television viewing that was responsible for the great drop in reading and mathematics scores. It was difficult to believe that the children in Illinois, Ohio, and New York watched television so much less than those living in California, Washington, and Idaho.

Certainly the ghettos of Chicago, Detroit, and New York had suffered a much greater social upheaval than most of the West, except for Los Angeles, and they clearly contained the greater number of minority groups, broken homes, one-parent families, and run-down schools plagued by vandalism, absenteeism, and violence in the classrooms. But the decline in scores did fit the pattern of the weapons-test fallout during the years of 1956 to 1958, when these children were born.

By far the largest amount of fallout from the Pacific and Siberian tests had rained out over Hawaii, Alaska, Washington, Oregon, and California, where the coastal mountain ranges near the population centers of Seattle and Portland showed the highest amounts of strontium 90, cesium 137, and other radioactive substances in the milk and diet. And in Nevada and Utah, the large 1957 series of tactical-nuclear-weapons tests had brought down the highest levels of short-lived iodine 131 in the milk ever recorded. It was, in fact, in the clean coastal counties of the Pacific Northwest where lung cancer rates had risen most sharply.

By contrast, the areas of the Midwest such as lower Illinois, Indiana, and Ohio had been spared the heaviest fresh fallout, which drifted mainly across the northern United States from the Nevada test site over Idaho, Montana, North Dakota, Minnesota, Wisconsin, and Michigan, until it reached the Appalachian Mountains of northern New York and New England. There the heavy rain and snowfall as well as the greater air pollution would bring it down in large amounts, just as the acid rain brought down the sulphur dioxide emitted by the coal plants of the Midwest. Together with the effects of poverty and ordinary air pollution of the industrial East, this could explain why the Middle Atlantic and New England states showed an intermediate drop in scores of only 12 to 14 points, between the West's 19 points and the 8 points of the Midwestern plain.

Bell and I realized that it would be extremely important to obtain the scores for some of the individual states, but these were not listed in any of the existing publications. Not until the summer of 1979, shortly before we were scheduled to present an invited paper at the meeting of the American Psychological Association in September, did we find a way to obtain the scores for four out of the five states for which the U.S. Public Health Service had measured the fallout levels back in 1957. Apparently, some of the states did not want their S.A.T. scores to be published, and so the College Board was able to release to us only the data for California, Utah, Ohio, and New York, but not for Missouri.

But as soon as I opened the letter with the data I knew that the hypothesis was once again supported by the evidence: By far the greatest drop between 1974 and 1976 had indeed occurred in the state with the highest levels of radio-iodine in the milk, namely Utah, and the smallest drop was recorded for the midwestern state of Ohio, largely to the south of the drifting clouds of fallout that had passed over Minnesota, Michigan, New York, southern Ontario Province in Canada and northern New England. The magnitude of the effect was difficult to believe, but here in the letter from the College Board were the hard numbers: Utah had dropped 26 points and Ohio only 2.

There was just no way that such an enormous difference in the sudden drop could be explained solely by socio-economic factors, differences in the quality of teachers, school curricula, television viewing, amount of cigarette smoking, drug use, alcohol consumption, or other gradually changing physical factors in the environment such as air pollution or pesticides.

In fact, if smoking, alcohol, and drug taking during pregnancy had been a factor, Utah, with its large Mormon population, should have declined less and not more than Ohio and New York. But it was the other way around: The population with lower cigarette consumption, alcohol, and drug problems during pregnancy had the greater decline in Scholastic Aptitude scores by many times the normal statistical fluctuation of 2 to 3 points.

Nor could differences in the genetic factors of the two populations be blamed: They were both predominantly white, and in fact the Mormons had originally come from the East and Midwest. Besides, genetic or inherited factors would lead to long-term differences, not the sudden changes that had taken place. Tragically, it now appears that we had unwittingly carried out an experiment with ourselves as guinea pigs on a worldwide scale. This discovery made me more determined than ever to do everything in my power to make sure that the terribly costly lesson would be learned before mankind would make further and perhaps more irreversible mistakes with fallout from nuclear war or nuclear reactor accidents, in which the radioactivity equivalent to a thousand Hiroshima bombs might suddenly be released over vast areas the size of entire states or nations.

But as we pointed out at the Psychological Association meeting, there was also reason for hope in the data for individual states. First of all, the test scores in Utah rebounded partially by 9 points the following year, when the nuclear test ban apparently showed its effect for the children conceived eighteen years earlier. At least some of the damage was not permanent, presumably because it affected the fetal thyroid more than the mother's, and the iodine was gone from the milk within a matter of a few months after the bomb tests ended, although damage from strontium 90 and its daughter product, yttrium 90, to the pituitary gland would continue for many years, since it had accumulated in the bones of young women for decades.

Also, it was encouraging that the average level of performance on the tests had been so much higher in Utah than in any of the three other states for which we had the data. Although California, New York, and Ohio all showed scores in 1974 that were above the U.S. average of 440, down 38 points from the maximum of 478 in 1963, Utah had by far the highest score, namely 532, compared with 459 for Ohio, 454 for New York, and 450 for California.

Thus, it appeared that given the kind of quality school system that the people of Utah had established, together with the good diet, the low amount of smoking, drug use, and alcohol consumption during pregnancy, it was possible to attain a much higher degree of performance on this type of achievement test. The potential for raising the school performance of our children in the future was therefore clearly immense if we could only learn to use our vast productive capability to provide better diets, better schools, and better prenatal care as the Mormons had been able to do. Instead we were investing more and more of our national income in gigantic nuclear reactors for both military and civilian purposes that were filling the air and drinking water with invisible radioactive poisons, destroying the most important resource of our nation, the physical and mental health of our children.

Furthermore, this high intellectual performance was achieved in Utah despite serious local pollution problems from copper smelters, large coal-burning plants, and as many automobiles per capita as anywhere else in the United States. (Neither copper smelters, coal plants, nor automobiles produce strontium 90 or iodine 131.) It was obviously not necessary to return to a primitive, nonindustrial society for people to have capable children or to live a long, healthy, and useful life. For not only did the people of Utah have children with test scores that were far above the average of the rest of the United States, but they also had among the lowest rates of heart disease and cancer in the entire nation.

Since the Mormon customs discouraged smoking, they did not experience the large synergistic multiplying factor for lung cancer that smokers experienced when the radioactive fallout arrived. Just as the uranium miners who did not smoke were better off than their coworkers who did, the Mormons were able to clear out of their lungs the fine radioactive dust particles more rapidly than those whose lung clearance was slowed down by the nicotine in cigarette smoke. Thus, their religious customs resulted in much lower amounts of both man-made and natural radioactivity staying in their lungs or entering their bloodstream, reducing greatly the risk of low-level exposure leading to rises in lung cancer, heart ailments, and other chronic diseases.

Both Bell and I were surprised by how well our rather startling hypothesis was received by the large number of psychologists who came prepared to question our theory. In the ensuing discussion, questions were raised as to whether a change in the mix of the students taking the tests might not explain a good portion of the long decline in test scores. That is, could the decline in scores be explained by the increase of students from lower-income homes who in the past would not have thought of going to college? This was indeed likely to be the case in the early years, according to some of the studies published by the Wirtz Commission, but it could not explain either the sudden sharp drop followed by a halt in the late 1970s, when the total number of students taking the tests was actually declining. Nor could an increase in the number of less well-prepared students explain the recent wholesale decline in the number of students who could score above 600 or 700 out of the possible maximum of 800 points in these tests.

It was, in fact, the extremely sharp decline in the number of very high-scoring students that presented the greatest potential problem for a society increasingly dependent on verbal and mathematical skills to run the computers, design the automated machines for the factories and farms of the future, administer an increasingly high-technology society, and operate the sophisticated electronic weapons of a modern army. Instead of 189,300 students who had been able to score above 600 in the verbal test among those born in 1952–53, there were suddenly only 110,300 for the birth years of 1957–58, a drop of 42 percent. And an even greater drop occurred for the top students on whom our society would depend for much of its new ideas, creativity, and leadership skills in the arts, the sciences, and engineering, namely those who were able to score over 700. In this category, the numbers were cut by more than half, from a high point of 33,200 born before the Nevada tests began in 1949–50 to a low of only 14,800 for those born in 1957–58, the years of the heaviest fallout from our weapons testing.

It was only too evident that if the radioactivity in the environment led to early infant mortality, childhood cancer, thyroid damage, and underweight births, then also the learning ability of the surviving children might never develop its full potential.

And it would be the steady decline in the ability to read and reason and not so much the rising cancer rates in old age that would be the real seed for the self-destruction of a modern technological society. The children that could not read or cope with mathematics and science would drop out of school and become permanently unemployable. And these young people would feel increasingly resentful toward those whom they blamed for their failure: their teachers, their parents, and their political leaders. Even worse, they would blame themselves and suffer from low self-esteem.

Many of the unemployed and discouraged young people would drift into crime, vastly raising the level of violence and fear in the cities. Not knowing what caused their problems, they would increasingly resort to drugs and alcohol to overcome their sense of failure and hopelessness, raising the rate of juvenile suicide and crime still further.

Not being aware of the subtle thyroid damage with its resultant lethargy, parents would blame the teachers, and teachers would blame the parents for the increasing loss of interest, discipline, reading ability, and general motivation of the students. Vast sums of money would be spent in efforts to help the slow learners and the many handicapped students suddenly flooding the schools, draining the resources of society at the very time when there would not be enough highly skilled, resourceful, and inventive young people produced to improve the teaching and raise the productivity of factories, businesses, and farms. At the same time, the cost of health care would spiral as more and more developed early chronic disabilities, a situation that would lead to increasing absenteeism from offices, schools, and factories, and thus further reduce the output of goods and services while expectations continued to rise.

As productivity dropped while the need for costly special education and disability payments rose, the vast amount of borrowing that government would have to do to provide for the rapidly growing number of unemployed, handicapped, and sick would drive up the rate of inflation more and more. To keep ahead of the inflation, as well as to dampen its flames, the banks would have to raise their interest rates so as not to lose money by lending. Industrial machinery could not be modernized because borrowing the money would become too costly. The factories and farms would fall still further behind in their ability to meet the growing demand for manufactured goods and food, further adding to the pressures of inflation.

At the same time, the smaller supply of capable and creative young people needed to fill the jobs as engineers, scientists, doctors, nurses, computer specialists, teachers, managers, and officers for the increasingly sophisticated factories, offices, schools, hospitals, and military services would drive up salaries, adding still more fuel to the inflationary fires. More and more plants would be forced to shut down because they could not compete with more modern factories in other countries whose young workers were more productive because these countries were not in the direct path of the fresh fallout from Nevada and therefore less heavily exposed to short-lived iodine. Also a greater fraction of the reduced supply of talented and inventive young people would be absorbed in the unproductive tasks of developing ever more complex and costly nuclear-weapons systems and reactors, thus further weakening the economic situation of the nation as it was forced to import ever larger amounts of civilian goods and machinery from other countries.

As I thought about this scenario, I wondered how much of this had already begun to happen, as juvenile crime and suicide suddenly doubled and tripled in the mid-1970s among the children born in the late 1950s all over the U.S. and in northern industrial countries, where the fresh fallout had come down most heavily. The end of weapons testing in Nevada had led to a halt in the decline of intellectual ability among those tested eighteen years later, especially those born well after 1963, when bomb testing ended. There were now fewer children born blind and deaf showing up in the statistics, and there were fewer leukemia cases, brain tumors, and suicides among children and adolescents. Fewer crimes were being committed by young people under 18 years old than during the mid-1970s, when the intellectual achievement scores had dropped most rapidly, although the latest crime statistics showed a second large jump in 1979, corresponding to the second series of heavy atmospheric bomb tests 17 to 18 years earlier in 1961–1962.

There were now also fewer who were born immature, underweight, and thus dying of chronic and infectious diseases, except near the growing number of nuclear reactors that started operating in the early 1970s.

There was to be yet another development that strongly supported the hypothesis that fallout had unanticipated effects on mental development of the young. Just six months after the meeting of the American Psychological Association in New York where we had presented our findings, another scientific meeting took place in Baltimore devoted to the biological effects of ionizing radiation. At this meeting, Dr. Charlotte Silverman of the Bureau of Radiological Health in the U.S. Department of Health and Human Services presented a paper entitled "Mental Function Following Scalp X-Irradiation for Tinea Capitatis in Childhood," a condition more commonly known as ringworm of the scalp.

Dr. Silverman summarized the results of studies of two groups of children treated by means of X-rays, a method no longer used. One group of 2,215 children was followed at New York University Medical Center, and another group of 10,842 children at the Chaim Sheba Medical Center in Israel was followed over a period of 20 to 25 years, together with matched groups of non-irradiated controls. Aside from an increase in the number of brain and thyroid tumors, there was also an excess of nervous, mental and behavioral problems in the irradiated groups. As Dr. Silverman reported, "The New York investigators found a higher incidence of treated psychiatric disorders among the irradiated which persisted during an observation time of about 30 years."

For the Israeli group as originally reported by Drs. B.

Modan and E. Ron at the Sixth International Congress of Radiation Research in Tokyo earlier that year, Dr. Silverman summarized the results as follows:

> Several measures of brain function, mental ability and scholastic achievement demonstrate that the irradiated children suffered impairment. These findings are consistent with and extend previous findings of suggestive brain damage from radiation.

The doses to the thyroids of the children were listed as having been in the range of 6 to 9 rads, well below the doses of 10 to 60 rads received by the children of Utah from the fallout of the Nevada tests reported by Dr. C.W. Mays at the August 1963 Congressional Hearings for the children of Utah in 1962, or the 5 to 40 rads estimated by Dr. Eric Reiss for the children of the Troy-Albany area where heavy rainouts in distant areas had first been discovered.

Since the thyroid doses to the more sensitive developing fetus are generally 10 to 20 times as great as for the young infant, it was therefore not surprising that effects on brain function, mental ability and scholastic achievement should be observable for the children born in Utah and other areas subjected to bomb fallout during the years of nuclear weapons testing.

Shortly after the Baltimore meeting, the College Board sent me the SAT scores for 45 out of the 50 states up to the most recent testing period of 1978–79, 17 years after the second nuclear weapons series of tests in 1962. It was clear that this more detailed data would provide a crucial test of our prediction that there should be another sharp drop in the scores associated with the series of Nevada bomb tests, and that the greatest declines should again be observed for the western United States downwind from Nevada.

After going through the table, I saw that the answer was quickly apparent. Among all the states listed, Utah again showed the sharpest decline in the entire United States, 11 points in a single year. The declines diminished with distance away from Utah across the northern United States until they reached 5 points in New York, 3 points in Connecticut and only 1 point in Rhode Island.

How long would it take before the public would be able to learn of these facts? How long would it take before the damage that the governments deeply committed to nuclear technology for weapons and energy were continuing to inflict upon their own children would be ended? There were, of course, hopeful signs among those born in rural areas far from the nuclear plants in the years since the bomb fallout had stopped. Would any government leaders dominated by fear of foreign enemies be able to find the courage to carry out the epidemiological studies that had been called for so often in the past and admit the tragic errors that had been made? Or were we all helplessly lock-stepped on the road to the self-destruction that blind persistence in the course we had taken would surely bring to our nation and all those who had followed us in the frantic rush toward the false promise of unlimited power presented by the discovery of self-sustaining fission?

All this had enormous implications for the proposed new missile systems and the scenarios used to justify their need. If indeed the fallout from the bomb tests in Nevada was the principal new factor responsible for the unprecedented sudden drop in mental abilities among newborn in those tragic years, then the detonation of just two or three half-megaton warheads on the missile silos in Nevada and Utah would cause a drop equal to that observed during the seven years of small-tactical-weapons testing. What then would a massive strike of many thousands of such warheads, exploded near the surface on our missile silos in the West, mean for the future of our nation, even if not a single one of our cities were destroyed? Knowing the enormous sensitivity of the fetus in the mother's womb, was it really credible that any President of the United States could be blackmailed into not firing our missiles after such a hypothetical attack in the hope of "saving" the population in the cities of the east? And what would it profit the leaders of another country that had launched such a blow against us if within days, a massive cloud of drifting fallout would poison the air, the food, and water of their children for generations, even if not a single one of our missiles should ever reach their land?

17

Incident at
Three Mile Island

A FEW MINUTES after ten o'clock on Wednesday morning,
March 28, 1979, I was sitting in my study trying to understand
the full implications of what I had learned during a week-
long trial in Philadelphia. At issue in the trial was the degree
of genetic damage done to the soldiers who had been marched
under the highly radioactive clouds during the Nevada tests
back in the 1950s. Just as I was trying to sort out my thoughts
on the discovery of the large doses the men had received by
breathing in the dust and gases, the telephone rang.

It was a reporter calling from radio station WPLR in New
Haven, Connecticut, a station from which I had received a
number of calls over the past year, ever since I had testified
at a Congressional seminar on the increases in infant mortality
and cancer following large radioactive releases from the Millstone Nuclear Plant some 25 miles east of New Haven. The
reporter wanted to know my reaction to a news bulletin that
had just come over the wire, according to which a general
emergency had been announced at the Three Mile Island Nuclear Plant near Harrisburg, Pennsylvania.

According to the report, high radiation levels had been

measured inside the main containment building of Unit II, and most of the plant personnel who were not essential were being evacuated following the accidental release of radioactive steam from the primary loop of the reactor.

From the little information in the first wire-service bulletin, I could only guess that this might be the beginning of a potentially serious accident. Normally, the steam from the primary loop of a pressurized-water reactor contains relatively low levels of radioactivity, and if there was enough to force evacuation of all but the most essential people in the control room, there had to be major damage to fuel elements in the reactor core. And since the motion picture *The China Syndrome* had just opened, my thoughts immediately turned to the possibility that right here in Pennsylvania, where the first commercial nuclear reactor had been built with so much hope, we might also experience the first melt-down and catastrophic release of radioactive gases about which a growing number of concerned scientists had tried to warn the public for years.

The first thing I did was to call Tom Gerusky at the state's Bureau of Radiological Health in Harrisburg, but the line was busy. I then decided to call the local offices of the Associated Press and United Press International in Pittsburgh to see whether more recent bulletins contained any more detailed information. What I learned from these reports was not reassuring. Apparently, radiation levels inside the containment building were still rising, and there were fragmentary reports of radioactive gas releases taking place, leading to above-normal radiation readings near the site, located some 10 miles south of the city of Harrisburg on the Susquehanna River.

At the same time, spokesmen for the utility were being quoted as claiming that there was no serious problem, and certainly no need for nearby residents to evacuate the area.

Which way were the winds blowing? Were they blowing north, toward the densely populated metropolitan area? I called the U.S. Weather Bureau and learned that early in the morning the winds had been blowing generally south and southeast, toward the rural counties of York and Lancaster, at a relatively low velocity. This was of course relatively good news for the large population of Harrisburg, but not such good news for the people in York and Lancaster counties. The low wind speed would mean relatively high concentrations of gases in the air, which in turn might lead to potentially large inhalation doses—as large as those I had just calculated for the soldiers in Nevada. And the fine drizzle in the air would bring down radioactive iodines into the local pastures, thus presenting still another problem for the people of this heavily agricultural area even if a major melt-down of the reactor core should be averted.

The telephone rang again, and this time it was someone from the Mobilization for Survival in Philadelphia, asking me whether I would be willing to go to a press conference in Harrisburg the next day together with Dr. George Wald of Harvard University. The purpose of this conference would be to present an alternative source of information for the people in the area on the potential health hazards from the accident. (So far, the people in the area had received nothing more than the bland reassurances being offered by the utility and the spokesman for the Nuclear Regulatory Commission, the government organization formed from the old AEC when it was reorganized a few years ago).

The thought passed through my mind that by tomorrow, Harrisburg might not be a very healthy place to bring a lot of reporters together for a news conference, but I tentatively agreed to go, provided there were no further serious or unforeseen developments. In the meantime, I would have to try to gather as much information as possible about just what was going on in the reactor. I needed to have a clearer feeling for the nature of the danger that the people in Harrisburg were facing.

Again I tried reaching Gerusky's office in Harrisburg, but without success. Obviously, everyone in the world was trying to reach him or Margaret Reilly, the only people who had any firsthand information on radiation levels aside from the utility's own people—and the utility's people could not be trusted to give out the true data.

The telephone rang again, and this time it was KDKA-

TV, the Westinghouse station in Pittsburgh, wanting to interview me for their evening news program. They read me the latest wire stories, and the implications were getting increasingly more serious as the full extent of the accident began to emerge. Apparently a cooling pump had failed to function, and a series of events led to the opening of a safety valve that allowed large amounts of cooling water to escape into the main containment building. This sounded ominous, but it was all still very confused.

Later in the afternoon, a reporter from ABC-TV in New York called to say that he heard that I was going to go to Harrisburg the next day. Would I be able to bring a survey meter along so that there would be some way that they could get independent information on what the radioactive levels at different distances from the reactor really were? I told him that I would try to do so if I could manage to borrow one from our nuclear medicine group.

That evening, Frank Reynolds reported on the *ABC Evening News* that there had been a large release of steam from the reactor early in the morning, that the accident had actually begun at 4:00 A.M., and that state officials were very upset about not having been notified immediately. There were apparently radiation releases from the turbine building, and there was indeed some damage to fuel rods in the core, as I had deduced from the earlier reports.

There was only one reliable source of information that I knew I could trust, and this was Henry Kendall of the Union of Concerned Scientists at M.I.T. Kendall had been responsible for bringing out the true danger of a major accident and the inadequacy of the emergency core-cooling system for preventing a melt-down of the core that would lead to the dreaded "China Syndrome." This was an in-joke among nuclear engineers for the scenario in which a molten mass of uranium, plutonium, and fission products would melt its way through the steel reactor vessel and through the concrete foundation deep into the earth, "all the way to China," with the release of much of the one thousand Hiroshima bombs' worth of radioactivity into the air if the containment were to be ruptured.

I finally was able to reach Kendall late in the evening at the home of his mother, who was seriously ill. He filled me in on what he had been able to learn. Apparently twenty-two previous cases of defects in the valves and pumps of this type of reactor had been reported to the NRC in recent years, but nothing had been done to correct the problem. As a result of the loss of cooling water, much more fuel damage had apparently occurred than had been expected. But there was essentially no adequate instrumentation provided to allow one to analyze exactly what was happening inside the reactor core in this kind of major accident. The designers had simply assumed that it would never happen.

The emergency core-cooling system apparently had been put into effect, but from what Kendall was able to piece together at this time, the reactor was at least not on what he called "a main track toward a complete melt-down," though a melt-down was still possible.

What was certain was that radiation levels inside the reactor containment building had risen to the highly lethal level of 4000 rads per hour, enough to kill an adult in five minutes, so that according to his rough calculation, even on the outside of the thick concrete wall the levels might be as high as 4 rads per hour. This implied a great deal of damage to the fuel elements in the core, which must have at least begun to melt. Reports that he had received also talked about external gamma radiation doses accumulating at the rate of 1 millirad every hour a mile or so away, or some one hundred times the normal background rate.

Kendall ended up by telling me how he and his associates had just discovered that five years before, the NRC had learned of serious defects in the computer programs for calculating the design of these plants to survive earthquakes, but that the NRC had kept it quiet until his group had discovered it independently. And he added that the Rasmussen estimate of the risk of a major accident had been underestimated by at least two hundred times, and that it was more like one in one hundred per year rather than one in twenty thousand per plant. Using Kendall's figures with fifty plants operating, a potentially major accident like Three Mile Island could happen once every few years.

This information made it clear to me that evacuation of the people, and particularly pregnant women, living within a few miles of the reactor should have been ordered long before, since the total doses to internal organs from inhalation of the fission gases were likely to be ten to one hundred times greater than the external gamma dose levels Kendall had told me about. Just as in the case of the Albany-Troy incident years ago, where the external whole-body dose was only about 100 millirads over a period of ten weeks, it would be the doses to the thyroids of the infants and the unborn in their mother's womb that would be much greater and far more serious in their effects. Perhaps in a single day, thyroid doses to the unborn would reach the values of a few hundred to a few thousand millirads, equivalent to a series of abdominal X-rays, for which Dr. Alice Stewart's data had indicated as much as a doubling or tripling of the risk of leukemia and cancer for those in the early phases of development.

Yet on the radio and television news that evening, there were still the bland reassurances from the Metropolitan Edison Company officials who operated the reactor. According to the president of the company, Walter Creitz, the public was not in danger, no one was killed, and no one had been injured by the accident.

There were also the usual reassuring phrases by the public-relations people of the NRC, with their carefully chosen qualifying words. According to them, there was "no immediate danger to life." Put in this way, it was literally true; so far, there were no immediately lethal doses, and any infants in their mothers' wombs who were endangered would not die until many months or years later, while some types of chronic diseases and cancers would not show up for decades.

At six o'clock the next morning the telephone woke me, and I was afraid of the news that it might bring. But it turned out to be bad news of a totally different and unexpected kind. It was my mother in Buffalo, who said that she could not sleep all night because of severe abdominal pain: could I please come to see her right away? She had suffered a heart attack a year before and had been in poor health, but more recently she had been well enough to do without a companion. As a result, she was now alone.

I did not know what to do. I was scheduled to be in Harrisburg at noon, and I did not see how I could break that commitment to the many people who were in danger from the radioactive gases still leaking from the plant. So I told her that I would call my brother in New York, and that if he could fly up in the morning, I would come to Buffalo in the afternoon, directly from Harrisburg. In the meantime, she should call her neighbor, who had been very helpful in the past, and ask him to drive her to the emergency room at the hospital, where I would call her.

Fortunately, my brother was able to go to Buffalo at once, and I set out for the airport to catch the flight to Harrisburg after stopping at the hospital to pick up the survey meter that I had agreed to take along.

I had checked the morning news just before leaving the house to learn the latest status of the plant. Daniel Ford, who worked with Henry Kendall, appeared together with Walter Creitz on the *Today* show. Apparently radiation measurements indicated a lower release rate than the night before, although radioactivity had by now been detected as far as 16 miles away. The temperature of the reactor was being lowered, and Creitz talked confidently about pumping out the radioactive water from the containment building in the hope of putting the reactor back on line again in the not-too-distant future.

Ford indicated that for the moment the reactor seemed to have been stabilized, that the emergency core-cooling system had been turned off, but that the NRC felt that there were still serious problems in keeping the reactor under control. At least it seemed to me that there had been no major deterioration of the situation during the night. From the airport I had called the hospital in Buffalo and learned that my mother had been sent home with what appeared to be nothing more serious than a stomach flu, and so I decided to get on the plane together with a great many other passengers, some of whom must also have wondered about the wisdom of flying to Harrisburg that morning. But none of them could have had any inkling of the full extent of the quiet tragedy that had already begun as the radioactive gases silently seeped from the damaged reactor only a few thousand feet away from where we would land on that gray, drizzly day in Harrisburg.

Just before boarding the plane, I decided to recheck the survey meter that I had placed in my briefcase. It still read the normal level of slightly less than a hundredth of a millirad per hour, and the reading did not change detectably after we had reached our relatively low cruising altitude for the half-hour flight east to Harrisburg across the two- to three-thousand-foot ridges of the Appalachian Mountains.

Next to me sat a young woman who evidently became quite curious when I opened my briefcase to look at the dial of the survey meter. She asked me what I was doing. It turned out that she was a nurse from our hospital on her way to Harrisburg to attend a conference on emergency medical care, and to compound the strange coincidence, her husband worked as a nuclear engineer for Westinghouse. I explained to her that I was planning to measure for myself the radiation levels in the Harrisburg area at different distances from the stricken plant in order to have some idea as to the magnitude of the hazard that the continuing releases were posing for the people in the area, particularly for the unborn.

Soon it came time for landing, and once again I turned on the survey meter to see what the radiation levels were a few thousand feet in the air, a few miles northwest of the Three Mile Island plant. As we both watched with growing concern, the needle began to move up-scale, until when we were just a few hundred feet in the air over the river close to the end of the runway, the meter indicated a dose rate fifteen times what would be normal. There could be no doubt about it: Some thirty-six hours after the accident, large amounts of radioactive gases were still escaping from the reactor whose twin cooling towers loomed ominously only a mile or so away through the haze. Apparently, the wind had shifted and the invisible gases were now drifting northwestward—up the river and toward Harrisburg.

The plane was delayed and so I was late for the news conference scheduled for noon in the Friends' Meeting House in
downtown Harrisburg. This meant that there was no time to
check the radiation levels still closer to the plant. But a quick
measurement outside the airport terminal showed the readings
to be ten times their normal value, confirming the high reading
in the plane.

On the way into the city, I noted down the readings every
mile as the taxi driver read me the distances. Three miles from
the airport, the readings dropped to only three to four times
normal, but at 4 miles, they rose again to eight and nine times
their usual rate. This meant that there were hot spots, either
due to gas pockets or to fallout deposited on the ground in
the course of the past day and a half of releases.

The high readings could not be due to any direct gamma
rays penetrating the four-foot-thick concrete walls of the reactor's containment building, since they would have diminished
steadily and rapidly with the increasing distance. But they were
consistent with large gas releases now drifting toward downtown Harrisburg, where the readings were still three to four
times the normal rate as we approached the dome of the State
Capitol 12 miles from the airport.

The news conference was already in progress when I arrived.
There were a surprisingly large number of reporters with micro-
phones, tape recorders, and television cameras crowded into
the relatively small meeting room, with Dr. George Wald of
Harvard sitting at a table toward one end.

I apologized for being late, and then took out my survey
meter to measure the radiation rate in the room. The reading
was still three to four times normal, or essentially the same
as outside. Clearly, the walls of the building did not provide
any significant protection. Most likely, it was the gamma radiation from the radioactive gas that was by now at the same
level as outside the building. Even closing the windows would
have been futile at this point.

The intensity of the questions from the reporters reflected
the great concern that existed, and I felt acutely the great
difficulty of having to explain, without causing a panic, the
seriousness of the situation that already existed for the pregnant

women and infants. I explained that at the moment, the radiation levels were not serious enough for the normal, healthy adult as long as they would not increase because of further releases. Asked what I would recommend in the light of my knowledge of the situation, I said that at the very least, pregnant women and young children should be urged to leave the area within a few miles of the reactor because of the likelihood of continuing releases of radioactive iodine that would concentrate in the fetal thyroid as well as in that of the infants and young children.

By limiting my recommendation in this manner, I hoped that there would not be any sudden rush toward a mass evacuation of the whole population, which might cause serious traffic jams and accidents. I was primarily concerned with preventing panic, especially since according to my latest information, there was apparently no immediate threat of a complete melt-down. And since the greatest danger existed for the unborn and very young, at least they would not be exposed any further, although at that point I did not know whether most of the dose had already been received, or whether there would in fact be any further large releases.

I also urged that pregnant women and young children should not drink fresh milk or local water for the next few weeks, until detailed measurements could be carried out to determine the precise levels of radioactivity. The most immediate hazard was clearly from the inhalation of the fresh radioactive gases by expectant mothers, which would lead in a matter of hours to significant amounts of radioactive iodine transmitted through the blood stream to the placenta and from there to the developing infant's small thyroid gland.

When someone asked Dr. Wald whether the public should believe me or the spokesmen for the utility who had just reassured them that there was no danger, he answered that under such circumstances, one should always ask oneself who has the greater financial interest, the industry or the concerned scientist trying to warn the public. Under the present circumstances, he personally would tend not to accept the reassurances of the industry spokesmen and would tend to believe that there was indeed reason for deep concern, as I had indicated. There was no safe level of radiation, and the unborn and the young are clearly more vulnerable than adults.

The news conference broke up shortly thereafter, and a number of reporters wanted to have more details on my findings. Unfortunately, I felt under great pressure to get back to the airport so as not to miss the next flight to Pittsburgh with a connection to Buffalo. I was still deeply troubled about my mother's condition, and I could not stay very long to answer all the many difficult technical questions posed by the reporters.

I did manage to catch the afternoon flight, and as soon as I got off the plane in Pittsburgh I went to a telephone to call my mother's house. When I received no answer, I had a deep sense of foreboding, and immediately called the emergency room at the hospital. The nurse who answered told me to wait a minute, until she could get the doctor on duty, and a few moments later I learned that my mother had just died in the emergency room from what appeared to have been a sudden, massive rupture of the abdominal aorta. The doctor told me that my brother had been with her, and that it happened so suddenly that she lost consciousness instantly. There was no long period of concern or pain, and she passed away in my brother's arms. For this I was of course grateful, but it could not change the fact that suddenly my mother was gone, and I had made the decision not to be with her at her time of greatest need.

The next few days were like a nightmare, in which I was torn between my private grief and the demands of an outside world clamoring for advice and help in the face of the growing fear that the reactor at Harrisburg might still melt down. An unexplained bubble of hydrogen threatened the efforts to keep the core adequately cooled, and all through this period uncontrolled releases of radioactive gases continued to take place, despite frantic efforts to bring the situation under control.

The next day, while making the funeral arrangements for my mother in Buffalo, I learned that Governor Thornburgh had ordered the immediate evacuation of all pregnant women and children below school age from the area around Three Mile

Island. It would be too late for many, but at least some lives would be saved, and I was grateful that my efforts to warn the people of the area had not been totally in vain.

Even though I had told my secretary that I could not take any calls in Buffalo, it was impossible to stop them all. In a way, the sense of being needed kept me from giving in to the deep sense of loss, and the continuing demands of life probably helped me to overcome the period of deepest grief. My mother had been a pediatrician and obstetrician, and she had always been greatly concerned about my findings. Somehow I knew that she would have wanted me to help the people who were so terribly troubled, even during this period of greatest personal and family upheaval.

When a few days later I tried to reach Henry Kendall to fill in the gaps in my knowledge of what was happening in the stricken reactor, I learned that his mother had also died during that terrible week. And, just as in my own case, the enormous needs of the outside world seemed to have helped him through his period of great personal crisis. It was a week that would forever remain deeply etched in our memories, and those of hundreds of thousands living nearby, who would never forget the days when their world had so suddenly threatened to come to an end.

On April 4, 1979, exactly a week after the accident at Three Mile Island had begun, Congressional hearings were scheduled to take place in an effort to learn what the long-range health effects of the accident were likely to be. They had originally been planned by Representative Lester Brown, and during the previous weekend, I had been asked whether I would be able to come to Washington to testify. It was not an easy decision to accept the invitation so shortly after my mother's death, but I agreed to come at the urging of environmental groups. The environmentalists feared that otherwise only officials of government agencies would be testifying, and in the past these officials had been on record as denying the seriousness of low-level radiation exposures from weapons fallout and normal nuclear plant releases.

Two days before they were to begin, the hearings were

shifted to the Senate under the chairmanship of Senator Edward Kennedy, and the environmental groups were told that my testimony was no longer desired. It was clear that both within the nuclear industry and the government agencies charged with the promotion of nuclear energy, every effort would now be made to save the industry. Clearly, this required that there should be no evidence presented that would suggest the possibility that anyone would die as a result of the accident.

President Carter, himself a former nuclear engineer trained in Admiral Rickover's nuclear submarine service, had just flown to Harrisburg together with his wife in order to reassure the people that there was no serious danger either from the gases that continued to leak from the damaged plant, or from the hydrogen bubble that was still threatening a melt-down.

At that very time, lawsuits were underway by servicemen who had been deliberately exposed to the radioactive fallout from nuclear-bomb tests in the 1950s. The servicemen were seeking compensation for the leukemia and cancers that had shown up among them at many times the normal rate. Another lawsuit, which had just come to trial in Philadelphia the week before Three Mile Island, involved a petition filed in behalf of men who participated in military exercises at the Nevada Test Site. At issue was whether or not the government should be required to notify the men that they had a significant risk of genetic damage that could affect their decision to have children. And finally, hundreds of individual lawsuits had been filed against the government by residents of Nevada and Utah for leukemia and cancer cases resulting from the years of exposure to the fallout clouds from the tactical-weapons tests.

Under these circumstances the last thing either the industry or the government wanted was testimony that might set off still another flurry of potentially costly damage suits in the Harrisburg area by women who were pregnant, some of whom might have miscarriages or lose their babies at the time of birth. Even the possibility of one such suit could threaten the survival of the nuclear industry, already reeling from the shock of an accident that it had assured the public would be as unlikely as being hit by a meteor while walking on the street.

The hearings by the Kennedy Committee did indeed go exactly as the concerned environmental groups expected. One government witness after another sent to testify by the White House assured the public that among the two million people living within 50 miles, and the hundreds of thousands who would normally be expected to die of cancer, there might perhaps be one or at most a few extra cancer cases, clearly a totally undetectable and therefore insignificant number. And, of course, not a word was said about the much more likely effects on infant mortality.

Only Dr. K. Z. Morgan, who had been one of the members of the panel appointed by Governor Shapp to hear the evidence on possible health effects of the Shippingport plant six years earlier, expressed concern over the neglect of the beta radiation in the official estimates of the radiation dose. But before he had a chance to explain the significance of the hundredfold greater beta as compared to gamma dose for internal organs such as the thyroid gland, the hearings were quickly adjourned.

Clearly, the public would once again be misled by the combined efforts of the old Atomic Energy Commission scientists now working for the NRC and the Department of Energy following the second reorganization of the old AEC. Once again, they were joined by the Department of Health, Education and Welfare, as had happened during the period of heavy bomb testing. Ironically, the previously secret details of the effects of bomb testing were being released that very week as a result of a Freedom of Information request filed by the *Washington Post.*

As told by Bill Curry in an article that appeared on April 14, 1979,

> Officials involved in U.S. atomic bomb tests feared in 1965 that disclosures of a secret study linking leukemia to radioactive fallout from the bombs could jeopardize further testing and result in costly damage claims according to documents obtained by the *Washington Post.* That study, as well as a proposal to examine thyroid cancer rates in Utah, touched off a series of top-level meetings within the old Atomic Energy Commission over how to influence or change the two studies.

The article then went on to say,

> The documents also indicate that the Public Health Service, the nation's top health agency, which conducted the studies joined the AEC in reassuring the public about any possible danger from fallout.

Here then was the long bitter story emerging at last, just as it was being repeated—not in the case of fallout from nuclear-weapons tests carried out in the national interest at distant test sites in the Pacific and the Nevada desert, but in the case of invisible releases from peaceful nuclear reactors near the nation's cities, in the private interest of an industry spawned by the secret military atom.

Nearly 40,000 pages of files dealing with radiation revealed a disturbing story of deception perpetrated in the national interest. Not surprisingly, the full consequences of this deception for the nation's health were never adequately examined.

Reading the list of what Curry discovered made me realize something that I had only begun to suspect in recent years, namely that some individuals in the government knew long before I had stumbled upon it accidentally how serious the fallout from weapons testing really was. As early as 1959, a study found higher levels of radioactive strontium 90 in the bones of younger children in the fallout zone. And, as Curry added, "coincidentally a Utah state epidemiologist found this year that children living in the zone during the weapons testing had 2.5 times as much leukemia as children before and after the testing program." This was the study by Dr. Joseph L. Lyon, published in the *New England Journal of Medicine* just a few weeks before Three Mile Island.

But what shocked me even more was Curry's account of a much earlier government study suggesting a link between fallout and leukemia that was begun even before I had submitted my first article to *Science* dealing with this possibility, back in 1963. Apparently, a 1959–60 spurt in leukemia in the southwestern Utah counties of Washington and Iron had been noticed by Edward S. Weiss of the Public Health Service, and he had immediately suspected fallout. The study, which showed that the two counties experienced 9 more leukemia cases than the 19 statistically expected, was essentially completed by July 1965, when Weiss submitted it for publication in a Health Service journal. Curry reported in the *Post* what happened next:

By September 1 of that year, a copy of Weiss's paper had been sent to the AEC, as had the Public Health Service's proposal to test school children in southwest Utah for thyroid abnormalities.

The AEC discussed the two studies that morning. The same day, a White House science adviser called the Health Service to ask, "What would be the federal government's liability for any health problems found?"

By five that afternoon, a joint AEC Health Service-White House meeting was set for the next day—with three HEW lawyers present, an extraordinary sign of the legal problems the studies could cause.

At the meeting, AEC representatives criticized the leukemia studies and the proposed thyroid study. It was agreed they would submit suggestions for changes.

A week later, the AEC was ready with a proposed letter to the surgeon general, the head of the Public Health Service. Dwight A. Ink, then assistant general manager of the AEC, told his commissioners:

"Although we do not oppose developing further data in these areas (leukemia and thyroid abnormalities), *performance of the . . . studies will pose potential problems to the commission: adverse public reaction, lawsuits and jeopardizing the programs at the Nevada Test Site.*" [Italics added.]

Not only would the study have jeopardized the commission's program at the Nevada Test Site for using strings of hydrogen bombs to build a new Panama Canal and to test designs for anti-ballistic-missile warheads in the atmosphere, but as I learned later, it might also have endangered the ambitious program of rapidly building a whole new generation of gigantic nuclear reactors all over the nation, each ten times as large as Shippingport, which were about to be considered for licensing. Among these were to be the plants of Beaver Valley, Millstone, and Three Mile Island.

As Curry's story made clear, this was to be the end of the report that might have given the public and the scientific community a timely warning of the unexpected seriousness of the planned normal and accidental releases of low-level radiation before the enormous financial commitment to a trillion dollars' worth of nuclear plants had been made by the nation's utilities.

In fact, it was clearly no coincidence that at exactly this time, namely the years 1964 and 1965, the Johnson White House had ordered a twentyfold increase in the permissible levels of iodine 131 and strontium 90 in the milk before it needed to be withdrawn from the market. (This fact came to light in the course of hearings by the Joint Committee on Atomic Energy on Radiation Standards held in 1965.) And it was also the time when the Johnson administration had made a secret commitment to a major involvement of American armed forces in Vietnam, where tactical nuclear weapons might have to be threatened or used if the Chinese should enter the conflict, as they had in Korea. That was clearly not the time to alarm the American people about the possible risk of leukemia, thyroid disease, and congenital defects among newborn children from the clouds of radioactive fallout that were certain to drift back over the United States if these weapons were ever used.

As Curry's story made clear, the AEC was determined to prevent the publication of the Weiss study, which would of course have fully substantiated the concerns of scientists such as Linus Pauling, Barry Commoner, Eric Reiss, E. B. Lewis, Jack Schubert, Ralph Lapp, myself, and many others who had warned of the possible rise in congenital defects, thyroid cancer, and especially childhood leukemia only a few years earlier. But our concerns had largely ended with the signing of the test-ban treaty by Kennedy and Khrushchev in the fall of 1963, just before Kennedy was assassinated. The release of the Weiss study would clearly have evoked renewed opposition from the scientific community and the public to the vast military and civilian programs that were being planned by the Pentagon, the AEC, and the nuclear industry for the use of bombs to dig canals and for vastly increasing the radioactivity in the environment from the production of weapons and the routine releases from giant commercial nuclear power plants.

The next part of the story in the *Washington Post* was therefore the inevitable next step in a Greek tragedy that would eventually lead to Three Mile Island and the crisis that a stunned nation would face when the promised source of cheap, clean, and economical nuclear power to replace the imported oil would suddenly turn into a national nightmare on their television screens:

> The next day, Sept. 10, Ink sent to the surgeon general a critique containing criticisms of the study's scientific basis which were made public in January with the Weiss report. The letter did not, however, make any reference to the AEC's concerns about damage suits, adverse publicity or its effect on the testing program.
>
> Meanwhile, the Public Health Service was gearing up to announce the thyroid study and to disclose the leukemia study. Weiss' study was formally prepared and dated Sept. 14. Two days later, the thyroid study was announced, but there was no mention of the leukemia findings.
>
> One Health Service document suggests that the service itself may have even suppressed the study temporarily to avoid excessive press coverage of the thyroid study. "All of this interest," an official wrote of the congressional and press concern for fallout studies, "will be intensified if publication of the leukemia portion of the study occurs before the [thyroid] project begins."
>
> Earlier, the Health Service had decided to minimize any publicity of the thyroid study.
>
> The result was that the Weiss study was not released and in 1966 was still under review and revision. It was never released.

It was now clear what Surgeon General Jesse L. Steinfeld had referred to when he answered an inquiry from Representative William S. Moorehead back in 1969. Moorehead wanted to know what had happened to the promised large-scale epidemiological studies on thyroid cancer, leukemia, and congenital defects in relation to fallout radiation requested by Congressmen Holifield and Price after the August 1963 hearings on low-level radiation. Steinfeld had written that the feasibility studies for such a program led to a decision that "a national program was not indicated" and that "the feasibility studies were not published." Those were the studies of Edward S. Weiss, A Public Health Service Officer who had tried to protect the lives and health of the people of the United States in accordance with his professional oath.

And as inexorable as that fateful decision was to suppress the truth about the biological effects of the worldwide fallout from nuclear-weapons testing in the interest of national security, it would now be necessary for the government to keep from the people of this country and the rest of the world the truth about what I knew would surely happen in the wake of the drifting fallout clouds from Three Mile Island.

18

Too Little
Information
Too Late

THE TRUTH WAS more difficult to suppress this time than it
had been in the atmosphere of fear engendered in the cold
war of the McCarthy years, the Cuban missile crisis, and the
Vietnam war. Nonetheless, an attempt to keep the facts from
the people was clearly being made.

The same Freedom of Information Act that had made it
possible for the *Washington Post* to reveal how the truth about
the bomb tests in Nevada had been kept from the American
people was used only a few days later to provide the first clue
to what really happened at Three Mile Island.

On Monday, April 16, 1979, only two days after Bill Curry's
story had gone out over the wires, the *Pittsburgh Post-Gazette*
carried across the top of its front page a special report from
the Associated Press with excerpts of tape recordings of the
proceedings of the Nuclear Regulatory Commission during the
crisis at Three Mile Island. Strangely enough, as in the case
of the Watergate affair, it would be the private conversations
of top government officials recorded on magnetic tape and unexpectedly released to the news media that would provide the
crucial information.

My own awareness of the existence of the tapes had begun with a rather amusing phone call from a *Washington Post* reporter. He wanted me to comment on a not very complimentary, and very bitter and sarcastic remark made by one of the NRC staff members on what I was likely to say about the doses that would be received by the people near Three Mile Island during the first few days of crisis. When the excerpts of the NRC officials' comments appeared in the *Post-Gazette,* I saw immediately that my worst fears about the true magnitude of the radiation doses were likely to have been correct. I was sure that there would once again be a large rise in fetal deaths, congenital abnormalities, infant mortality, and childhood leukemia, followed by the delayed rises in infectious diseases, heart disease, and cancer among those exposed in the years to come.

The excerpts opened with the following comment by Lee V. Gossick, Executive Director for Operations of the NRC on Friday, March 30, the day after I had urged the evacuation of pregnant women and young children at the news conference in Harrisburg:

"Bill, we have got a deteriorating situation up there with regard to some releases. The Governor is asking us to confirm what he is getting from the plant, which says that they had an uncontrolled release of stuff."

To this, Harold Denton, Director of the Office of Nuclear Reactor Regulation in charge of all NRC activities at Three Mile Island, answered:

"They are getting 63 curies per second, and I can't explain to you the mathematics, but what they are saying is if that's true, by comparing it with what we know the shutdown rate was and the measurements taken at the north gate, and those were yesterday, they were on the order of three times what they were yesterday, which would put us in the 1200 millirems per hour."

These were truly enormous release rates of fresh fission gases, since in a single hour (consisting of 3600 seconds) there would be 3600 times 63 or some 226,000 curies being released in an uncontrolled manner without detailed analysis or significant hold-up to allow the most dangerous short-lived isotopes to decay. Even Millstone, the worst of all reactors, had not released more than fifteen times that much in a whole year.

The corresponding radiation dose of 1200 millirems per hour, equivalent to the dose of some 50 to 100 chest X-rays, was presumably measured in the narrow plume close to the release point at the plant, since it was ten thousand times the dose-rate that I had recorded the previous day in the airplane and on the ground a few miles away in the pockets of trapped gas. The road into Harrisburg was evidently well away from the narrow, invisible plume of radioactive gases meandering upstream, to the northwest, toward Middletown and Harrisburg with the slowly moving prevailing winds. Wherever it touched down, it would lead to an enormous inhalation dose in a very short time.

The taped conversation continued with the following question by Peter Bradford, one of the NRC commissioners who was not trained as a nuclear engineer:

BRADFORD: What actual measurements do you have?

GOSSICK: I can't give you any at the moment. I don't have anything that is current since this happened here, you understand. The source has been sealed up again and I think this is probably being released for one or two hours. We don't know, however, whether that's good for any period of time.

GILINSKY (NRC Commissioner): Do we have any monitoring equipment—

DENTON: There is a lot up there, Vic, but it takes a while for it to ever get back here.

FOUCHARD (Director of Public Affairs for the NRC): I just had a call from my guy in the Governor's office and he says the Governor says the information he is getting from the plant is ambiguous, that he needs some recommendations from the NRC.

DENTON: It is really difficult to get the data. We seem to get it after the fact. They opened the valves this morning, on the let-down, and were releasing at a six-curie-per-second rate before anyone knew about it. By the time we got fully up to speed, apparently they had stopped, there was a possible release on the order of an hour or an hour and a half—

GILINSKY: And when did this plume—when was the puff released?

DENTON: Within the last two hours.

HENDRIE (Chairman of the NRC): Presumably it has just terminated recently, then.

DENTON: We don't know how long, but if it was a continuous release over a period of an hour or an hour and a half, which from what I understand which is a kind of lot of puff.

HENDRIE: A couple-of-knots wind and the dammed thing—the head edge of it is already past the five-mile line.

HENDRIE: There has been a suggestion for a five-mile evacuation in the northeast direction. I take it—

DENTON: A good five miles, I would say from first impression.

BRADFORD: It ought to made clear that you are not talking about lethal doses.

FOUCHARD: Mr. Chairman, I think you should call Governor Thornburgh and tell him what we know. I don't know whether you are prepared at the present time to make a commission recommendation or not. The Civil Defense people up there say that our state programs people have advised evacuation out to five miles in the direction of the plume. I believe that the commission has to communicate with the Governor and do it very promptly.

GILINSKY: Well, one thing we have got to do is get better data.

FOUCHARD: Don't you think, as a precautionary measure, there should be some evacuation?

HENDRIE: Probably, but I must say, it is operating totally in the blind and I don't have any confidence at all that if we order an evacuation of people from a place, where they have already gotten a piece of the dose they are going to get into an area where they will have had 0.0

of what they were going to get and now they move some place else and get 1.0.

GILINSKY: Does it make sense that they have to continue recurrent releases at this time?

DENTON: I guess I tend to feel that if they really didn't stop the release a half an hour ago—it's probably best to leave it to the operational people up there. The cloud hasn't had a chance to get down to these low levels.

AHEARNE (NRC commissioner): But Harold, what confidence do you have that they won't embark on the same thing?

DENTON: I don't have any basis for believing that it might not happen—is not likely to happen again. I don't understand the reason for this one yet.

HENDRIE: It seems to me that I have got to call the Governor . . . to do it immediately. We are operating almost totally in the blind, his information is ambiguous, mine is nonexistent, and—I don't know, it's like a couple of blind men staggering around making decisions.

This last remark by Chairman Joseph M. Hendrie was to be quoted repeatedly by the news media later, but strangely enough it was always referred to in the context of the danger of a major melt-down. But clearly, either by reason of ignorance or design, the public was not told that the remark was made in the context of the danger from the ongoing large gas releases that were rapidly approaching the critical EPA emergency dose of 25,000 millirems to the whole body or key organs such as the thyroid from internal and external sources combined. This high level had been set by the Federal Radiation Council back in 1965 and later adopted by the EPA as the maximum allowable dose before countermeasures or an evacuation of the population should be ordered. Certainly neither the people living nearby nor the reporters at the site had ever been told of the large radiation doses from the inhaled gases that they were receiving.

The excerpts from the tapes continued to record the enormous concern about the growing radiation doses from the inhalation of the gases to the people in the area:

> FOUCHARD: Is there anybody who disagrees that we ought to advise the Governor on what to do?
>
> DENTON: I don't. Just on the basis of what we know. It's a good first step.
>
> HENDRIE: Go ahead with the evacuation?
>
> DENTON: I certainly recommended we do it when we first got the word, commissioner. Since the rains have stopped and the plume is going—I would still recommend a precautionary evacuation in front and under. And if it turns out we have been too conservative—
>
> GRIMES: My view is that it might have been useful right near the site, but now it is down below the EPA [unintelligible] level, so it probably is the most that should be done, in my view, is to tell people to stay inside this morning.
>
> AHEARNE: I was going to ask, what about pregnant women and children?
>
> GILINSKY: Well, Brian says it is a factor of 10 that can be gained by staying indoors. Anyway, I just think it is worth getting that half hour to find out, first of all, you are alerting people that they are going to have to do something, and they are not going to be able to do something in a half-hour anyway.
>
> HENDRIE (on telephone): Governor Thornburgh, glad to get in touch with you at last. I must say that the state of our information is not much better than I understand yours is. It appears to us that it would be desirable to suggest that people out in that northeast quadrant within five miles of the plant stay indoors for the next half hour.
>
> We have got one of those monitoring aircraft up and seem to have an open line to it and we ought to be able to get some information in the next 10 to 15 minutes. They can tell us whether it would be prudent to go ahead and start an evacuation out in that direction.

THORNBURGH: So your immediate recommendation would be for people to stay indoors?

HENDRIE: Yes, out in that—out in the northeast direction from the plant.

THORNBURGH: The northeast direction from the plant to a distance of?

HENDRIE: To a distance of about five miles.

I have got a reading. During one of these burst, releases up over the plant several hours ago, up over the plant about 1200 millirems per hour which seems to calculate out, by the time the plume comes to the ground where people would get it, would be about 120 millirems per hour. Now, that is still below the EPA evacuation trigger levels; on the other hand, it certainly is a pretty husky dose rate to be having off-site.

Here it was: the NRC knew that the true doses were not just a few millirems to the people in the area, as had been claimed at the Kennedy hearings, with maximum values of the order of 75 to 100 millirems nearest to the plant.

They knew just as I did that the greatest dose arose not from the external gamma radiation measured by a survey meter or a film badge, but from the internal beta radiation from the inhaled fission gases and particles in the lung, the thyroid, and the other critical organs that concentrate the different substances according to their various chemical properties. So when the external gamma-dose rate on the ground was of the order of 1 to 2 millirems per hour, the true dose rate to the lung and other critical organs could be as much as 50 to 100 times greater, or of the same general magnitude as the 120 millirems per hour Hendrie himself had just mentioned.

But they also knew, as I did, that if they ever were to order the full evacuation that should have been ordered long before, it would not only have caused a panic among the completely confused and unprepared population, it would also have been the end of the nuclear industry, whether or not the core would ever go to a complete melt-down.

And so more precious time in which to save lives was being

lost by recommending only that people should stay indoors, as was clear from the taped conversation with Governor Thornburgh. At that very moment Governor Thornburgh was under enormous pressures from those wanting to protect this important Pennsylvania industry on the one hand, and his Secretary of Health, Gordon MacLeod, on the other hand, who was at that very time urging that at least the pregnant women and young children should be evacuated.

But the NRC was clinging to the hope that an evacuation of any kind would not have to be ordered because of the continuing gas releases:

KENNEDY: Don't we know that it has been stopped?

FOURCHARD: Vice-President Herbein of the company reports it has been stopped. Chairman Hendrie has talked to the Governor and recommended that he advise people to stay indoors up to a distance of about five miles for the present time.

GOSSICK: We have just lost telephone contact with the site. I assume that it is telephone problems, but—[inaudible]. Okay, we have got communications with a trailer up there, but we have lost contact with the control room. . . .

GILINSKY: Let me ask you, what is the status of the reactor, by the way?

CASE: Same as it was an hour ago, I guess.

GILINSKY: What is the state of the core?

CASE: Well, it is about like it was yesterday. The temperature is about 280 degrees, the pressure is up [inaudible] . . .

Here was the proof that up to that very moment the primary consideration in the decision to evacuate the people was not the awareness of a serious chance of a melt-down of the core, but the continuing uncontrolled releases of radioactive gases that were threatening to give internal exposures to the public on the order of tens of thousands of millirems.

In fact, the subsequent excerpts made it clear that only

at this point in the desperate deliberations did the knowledge of extensive core damage and a hydrogen explosion reach the NRC commissioners:

HENDRIE (TO DENTON): I have talked to the President and I think you ought to go down to the site. He will be sending down, immediately, a sort of communication system that he takes with him when he travels. He will send somebody with it and he wants to be in a position to pick up the telephone and go right through to the site, and be able to talk to his man down there for information and recommendations on what to do. . . .

MATTSON: [Director of the Division of Systems Safety] . . . Now, B and W [Babcock and Wilcox] and we have both concluded . . . that we have extensive damage to this core. That corroborates with the releases we are seeing. . . . My best guess is that the core uncovered, stayed uncovered for a long period of time, we saw failure modes, the likes of which has never been analyzed. . . . We just learned . . . that on the afternoon of the first day, some 10 hours into the transient, there was a 28-pound containment pressure spike. We are guessing that may have been a hydrogen explosion. They, for some reason, never reported it here until this morning. That would have given us a clue hours ago that the thermocouples were right and we had a partially disassembled core.

HENDRIE: Where abouts is the bubble?

MATTSON: The bubble is in the upper head. The upper head volume at 1128 cubic feet as best we can tell. The estimate of the gas in that volume now is 1000 cubic feet, best that we can tell. That is at 1000 psi. If you take the plant to 200 psi, then—

HENDRIE: Yes, you are going to blow right down and empty the core.

MATTSON: I have got a horse race. I'm putting in high head, and if I get down in pressure, low head and coolant, it is coming in the cold leg, it is going down to the lower plenum, it is coming up through the core, it is splashing and it runs into the noncondensibles, I've got a core partially full or maybe totally full of noncondensibles. . . . We have got every systems engineer we can find, except the ones we put on the helicopter, thinking the problem, how the hell do we get the noncondensibles out of there? . . . Do we win the horse race or do we lose the horse race? And if you are lucky and there is not a lot of—you have overestimated the noncondensibles, you might win. If you are not lucky and you have got the right number on the noncondensibles you might lose it.

HENDRIE: . . . It sounds to me like we ought to stay where we are. I don't like the sound of depressurizing and letting that bubble creep down into the core.

And then came the following most revealing words from Roger Mattson, Director of the Division of Systems Safety, whom I knew to be very knowledgeable with regard to radiation dose calculations.

MATTSON: Not yet. I don't think we want to depressurize yet. The latest burst didn't hurt many people. I'm not sure why you are not moving people. Got to say it. I have been saying it down here. I don't know what we are protecting at this point. I think we ought to be moving people.

HENDRIE: How far out?

MATTSON: I would get them downwind, and unfortunately the wind is still meandering, but at these dose levels that is probably not bad because it is [inaudible].

KENNEDY: But downwind how far?

MATTSON: I might add, you aren't going to kill any people out to 10 miles. There aren't that many people and these people have been—they have had two days to get ready and prepare.

KENNEDY: Ten miles is Harrisburg.

MATTSON: 40,000 [inaudible] five miles. . . . It's too little information too late, unfortunately, and it is the same way every partial core melt-down has gone. People haven't believed the instrumentation as they went along. It took us until midnight last night to convince anybody that those goddamn temperature measurements meant something. By four o'clock this morning, B and W agreed.

Not until later in the day did Governor Thornburgh finally agree to the compromise plan of ordering evacuation of pregnant women and young children. But as I was to learn in an unexpected manner a few months later, most of the damage had already been done before the evacuation was ordered.

During the latter part of the summer I received a phone call from someone who asked me for a collection of all my articles on the effects of low-level radiation. He said that he was working on a study of the total economic and health impact of the Three Mile Island accident for the Presidential Commission chaired by Dr. George Kemeny of Dartmouth College, and he wanted to include an upper-limit estimate based on my statistical findings around various nuclear plants and after various fallout episodes.

I was surprised by this request, since I had not been asked to testify before the Kemeny Commission, which had been holding extensive hearings all summer long. Also, I wondered how the caller could use my earlier studies if there were no detailed estimates of releases and dose measurements available. The NRC had claimed that all the meters in the stack had gone off scale so that no one knew or would ever know just how much had actually been released.

And so I asked whether he had any detailed information on the quantities of gases released or the amounts of radioactivity in the air, the milk, and the diet on which to base a meaningful estimate of the likely health effects and their costs.

To my surprise, he said that he had such a document prepared for the Metropolitan Edison Company by their own environmental consultants, Pickard, Lowe and Garrick, and that he would be glad to send me a copy for my examination, since it was being widely circulated within the NRC and the Kemeny Commission.

A few days later, the two-inch-thick document entitled "Assessment of Offsite Radiation Doses from the Three Mile Island Unit 2 Accident TDR-TMI-116," dated July 31, 1979, arrived in the mail. And there, in the second paragraph, was the proof that the evacuation ordered on the third day had indeed been too late:

> Based on techniques used in this analysis, dose estimates are consistent with the release of seven million curies of noble gases in the first one-and-one-half days of the accident, two million in the next two days and one million in the next three days, and a relatively small amount thereafter.

By Friday afternoon, the third day of the accident, when the evacuation took place, between 7 and 9 million curies out of an estimated 10 million curies of radioactive fission gases had already been released, together with a corresponding fraction of the 14 curies of radioactive iodine 131 as given on page III of the report. And since the report also concluded that most of the thyroid dose was due to inhalation—and not ingestion of drinking water or milk—in the first five days of the accident, it was clear that by the time the evacuation of the pregnant women had been ordered, most of the thyroid dose to the developing fetuses had already taken place.

These were indeed very large amounts of radioactivity, comparable to those that arrived from the Chinese bomb tests in October 1976 on the East Coast, and for which I had found a 20 to 60 percent increase in infant mortality in the following three months all the way from Delaware to Maine. Just two years earlier, I had prepared a paper on this incident for the Committee on the Biological Effects of Ionizing Radiation of the National Academy of Sciences for its meeting in Washington on July 17, 1977, in which the levels of radioactive iodine 131 in the milk reported by the EPA were summarized.

Looking at the Metropolitan Edison Company's own measurements in some of the nearby towns, I saw it was evident that levels of iodine 131 had been produced by the Three Mile

Island accident comparable to those that had been produced by the heavy rainout of the drifting Chinese fallout as it met a severe rainstorm moving up the East Coast from Delaware to Maine. For the Hardison Farm, to the north of Three Mile Island, the report listed 110 picocuries per liter on April 25, almost a month after the accident. By that time, the activity would have decayed away to only about one-eighth to one-sixteenth of its maximum value, so that the peak values in local farms, if there had been adequate monitoring at the time, would have been in the range of 800 to 1600 picocuries per liter. The amount one worried about was measured in picocuries, which was only a millionth of a millionth, or a trillionth, of one curie. Yet 14 curies or 14 trillion of these units had been discharged into the air, breathed by the pregnant women in the area, and added to what they ingested with the milk. The report showed concentrations in the air as high as 20 picocuries per cubic meter. Since 1 cubic meter was roughly the volume of air inhaled by an adult every hour, 20 picocuries of iodine 131 entered a pregnant woman's lungs each hour at these concentrations.

The milk concentrations compared with the highest values listed in the EPA's December 1976 tabulation published two months after the Chinese fallout had arrived. These had ranged from 36 picocuries per liter in Rhode Island to 123 in Connecticut. At that time, infant mortality rose 60 percent in the first quarter of 1977 compared to the same period in 1976 in Delaware, 41 percent in New Hampshire, 17 percent in Maine, and 13 percent in Connecticut. But what was even more significant was that in Massachusetts, where the health department had ordered the cows to be fed stored hay which was relatively free from fresh fallout, infant mortality continued its rapid decline. Furthermore, infant mortality kept decreasing in the United States as a whole by 7 percent, as it did in Rhode Island, which receives most of its milk from Massachusetts.

Thus, if the measurements reported by Picker, Lowe and Garrick were accurate, there simply would have to be a sharp rise in infant mortality in the Harrisburg area and those parts of Pennsylvania and other nearby states over which the radioactive gases released in the first three days had drifted. On the other hand, infant mortality should again continue its normal decline in areas that happened to be spared by the invisible clouds of radioactive gases from the Three Mile Island nuclear plant.

But in which direction was the wind blowing during the period of highest release rates? Was there any way to find out how much was coming out at any given moment? After all, according to a story sent out by the *Washington Post* News Service and published on April 22, the Nuclear Regulatory Commission had been told by one of its staff people, Albert Gibson, that the radiation monitors in the stack went off scale on the morning of the accident. Thus, in answer to Commissioner Victor Gilinsky's question, "So we don't really know what went up," Gibson replied, "That's correct."

The story went on to say that as much as 365 millirems per hour of beta and gamma radiation were recorded on the ground some 1000 feet from the stack, and a helicopter had recorded three times this level in the air over the vent, confirming once again that the dose rates were far higher than the public had been told at the Senate hearings when the beta radiation that accounts for most of the internal dose from inhaled gases is taken into account. But Gibson went on to say that "those measurements were very inconclusive," and that "without knowing the precise weather patterns, we don't know if they were made at the appropriate locations."

However, leafing through the report of Met Ed's environmental consultants, I found that these were all completely misleading statements. Contrary to what the NRC commissioners and the public were being told, there were radiation monitoring instruments in the plant that never went off scale, namely in the auxiliary building, whose readings were directly related to the amount of radioactive gas being released. Here was the way the utility's consultants described how it was possible to know how much gas was being released every moment:

Strip chart records from all noble gas radiation monitors in the plant ventilation exhaust show no significant radiation levels during the first three hours of the accident. Since these monitors are in the most probable pathway for release, it is concluded that no significant releases occurred before 0700 March 28. Shortly after 0700, however, these monitors, which are designed to read normal low levels, indicated rapidly increasing radiation concentrations. Within a few minutes, they went off scale on the high side. At about the same time, the in-plant building area monitors which measure radiation levels inside the fuel handling and auxiliary buildings began to record increasing levels from about 1 milliroentgen to 100 milliroentgen per hour at 0740. At about 0900 the readings began to increase again to reach about 100 milliroentgen per hour at 1000 hours. They continued to fluctuate at high levels for about four days. One or more of these area monitors continued to read on scale during the course of the accident.

The report went on to explain in detail that by means of these measuring instruments it was possible to know what went out the stack because:

. . . radiation levels measured by area monitors in the auxiliary and fuel handling buildings are proportional to the rate at which airborne gamma activity was released to the environment . . .

In table after table and chart after chart, the releases and gamma radiation doses in different directions were worked out in detail. For every hour of the accident from 4 A.M. on March 28 until midnight of the fourth day, the readings of the area monitors were given together with the hourly wind direction and wind speeds. It showed that during the period of highest releases, from 10 A.M. on Wednesday the 28th to 7 A.M. on Thursday the 29th, the winds were blowing north, northwest, and west at 6 to 9 miles per hour, sending the radioactive gas toward upstate New York and western Pennsylvania. Only later, when the rate of release had decreased tenfold, did the winds shift briefly to the south, becoming more variable thereafter.

By the time the winds were blowing toward the northeast on Saturday, the fourth day of the accident, the intensity had dropped to less than one-twentieth of its peak value, thus largely sparing the most densely populated areas of Philadelphia, New Jersey, and New York City.

No wonder the NRC staff did not want to let the public know that they knew exactly in which direction the most radioactive clouds had moved, since this information could then be used to tie any later localized rises in fetal deaths, infant mortality, and cancer to the radioactive gas clouds from Three Mile Island. In fact, I remembered only too well the attempts of some of the same individuals formerly working for the AEC to discredit my findings on the rises of infant mortality across the southeastern United States following the first nuclear-bomb test at Alamogordo by claiming that the winds were not blowing in that direction. And they certainly did not want any of this to become known before the Kemeny Commission was scheduled to complete its report in early November.

Once again, as in the case of the Nevada tests, it was essential to keep such knowledge from the public and the scientific community at large. The NRC, the EPA, and all the other federal and state agencies knew full well that the doses were comparable with those experienced by the people of Utah, Montana, Wyoming, and the other states across the northern United States as far as New York and New England during the period of the Nevada tests, or for releases from some of the largest and most heavily emitting reactors, such as Millstone in New London, Connecticut, over a period of a year or two.

If, indeed, there should once again be sudden rises in infant mortality in areas where the radioactive clouds had drifted and the public should learn of them when the televised nightmare of Three Mile Island was still fresh on everyone's mind, this public knowledge would threaten the government's and the nuclear industry's vast program to build a thousand of these giant reactors by the end of the century near all the major cities, and would result in costly damage suits, exactly as in the case of the Nevada tests.

I had tried to obtain the preliminary data on monthly infant mortality rates by county from the Health Department in Harrisburg without success. I was told that this data had not yet been sorted out and processed, and that it would be many months before it could be properly assembled, even though such data had already been sent to me by the Maryland Health Department. All research on the health effects of Three Mile Island in Pennsylvania were under the direction of Dr. George Tokuhata, the same man who prepared the statistical portion of the Shapp Report exonerating Shippingport back in 1973. There was clearly no hope for any help from that direction.

All I could do was to wait for the state-by-state data on monthly infant deaths and births published by the Center for Health Statistics in Washington, which was usually three or four months behind. Thus, if significant effects would first show up for infants two to three months after the accident, or for the months of May and June, the earliest numbers indicating an effect would not become publicly available until August or September, too late for any presentation to the Kemeny Commission.

Nevertheless, I decided to gather whatever data I could as soon as possible, for an enormous media campaign had been launched by the nuclear industry to convince the public that there were no serious health effects due to the accident at Three Mile Island. No one had died, and no member of the public had been injured; the safety systems had worked, and there was no reason to abandon this important source of energy at a time when the United States depended so heavily on imported oil from the unstable Middle East.

In mid-August, the latest monthly report from the U.S. Center for Health Statistics for the month of May arrived in the library. Calculating the rates of infant deaths per 1000 live births, I found what I had expected. Instead of declining from the winter high, infant mortality in Pennsylvania had gone up following the accident at the end of March. Compared to 147 deaths in February and 141 in March, there had been 166 in April and 198 in May, an unprecedented rise of 40 percent. Yet, the number of births had actually declined from 13,589 in March to 13,201 in May. Thus the rate of infant deaths per 1000 live births had increased even more, namely by 44 percent, from 10.4 in March to 15.0 in May.

Yet, at the same time, the rate for the United States as a whole between March and May had declined 11 percent as it normally did, dropping from 14.1 to 12.6 per 1000 live births.

These were highly significant changes, the Pennsylvania figures for March and May representing an increase of 57 deaths, which was more than three times the statistically expected normal fluctuation of about ± 16, and thus unlikely to occur purely by chance in less than one in a thousand instances. But how else could I test the hypothesis that these increased rates were likely to be due to the releases from Three Mile Island without having the county-by-county and the Harrisburg figures available to me?

Having learned from the utility's own report that the heaviest releases had occurred when the wind was blowing north, northwest, and west, and having seen reports in the papers of high levels of radioactivity being measured in Syracuse, New York, some 150 miles to the north, at the time of the accident, I decided to examine the figures for New York State. By a fortunate coincidence, the U.S. Vital Statistics gave separate figures for New York City and the rest of the state, most of whose population was located in upstate New York—north, northwest, and northeast of Harrisburg some 100 to 200 miles away. Here, then, was a clear prediction of the hypothesis that could be tested: The figures for the rest of the state outside of New York City should have gone up, while New York City should either have shown no change or an actual decline.

And this is exactly what the numbers showed: Between March and May, infant deaths outside New York City climbed an amazing 52 percent, by 63 deaths, from 121 to 184. For New York City during the same period the number declined from 166 to 129. Again, these changes were many times as large as normal fluctuations, and the number of births changed relatively little, or by less than 10 percent, so that there could be no doubt about the significance of these changes in infant deaths.

So far, the hypothesis had passed its first major tests, but would it hold up for the other nearby states? What about Maryland to the south, where some of the gases had drifted in the morning and afternoon of the second day, according to Met Ed's report. The numbers were smaller than for the more populous states, but the changes continued to support the hypothesis: Infant mortality rose 26 percent, from 39 to 49 deaths, while the number of births remained essentially unchanged—4013 in March and 4076 in May.

What about New Jersey, to the east and northeast of Harrisburg? If the hypothesis was correct, there should not be any significant increase, since by the fourth day of the accident, when the winds shifted toward New Jersey, the rate of release had already sharply declined. Again, the hypothesis held up under the test. Between March and May, New Jersey infant mortality rose by only 8 deaths, from 87 to 95. This is not considered a significant increase, with the spontaneous statistical fluctuation of about ± 13 normally expected.

What about Ohio to the west, which had for decades closely paralleled Pennsylvania in its declining infant mortality figures? Did it show the same 40 to 50 percent rises of Pennsylvania and of New York State outside New York City? It was not to be expected that the gases would have drifted more than 200 to 300 miles west of Harrisburg, counter to the generally prevailing west-to-east movement of air masses across the United States, and so there should really be very little change in the Ohio figures. Again the numbers bore this out: Between March and May, infant deaths declined in Ohio from 177 to 160, the rate remaining constant at 11.5 per 1000 live births.

There simply could be no other explanation for such a localized pattern of sharply increased infant deaths in the areas where confined winds had blown the radioactive gases, while infant mortality rates were steady or declined in all the surrounding states that were not in the direction of the winds during the first two days of highest releases.

And yet, it would be important to have some figures for the area that was most heavily exposed close to the plant. If the figures around Harrisburg should indeed show much higher rises in infant mortality than the 44 percent for Pennsylvania and the 50 percent for New York State outside New York City, then it would be difficult to reject the hypothesis that it was indeed the radioactive gases from the stricken plant that were responsible for the unusual increase in newborn deaths.

But only Tokuhata had the data for the 5-mile and 10-mile zones around the plant, and there was no way that I would be able to obtain them. From everything I had been able to piece together, the numbers in Harrisburg had to show very large increases if areas as far away as 150 miles in upstate New York showed 50 percent rises in infant deaths. From my earlier studies on fallout clouds, I knew that the effect would roughly decrease in inverse proportion to the distance from the point of release. Thus, for Harrisburg, only 10 miles to the north of the plant, the rise in newborn infant mortality would have to be as high as 300 percent to 600 percent, corresponding to a four- to sevenfold increase above normal to be consistent with the rises in upstate New York.

How could I get at least an indication of whether this was the case? Just at this very moment, a way opened up to obtain this crucial information without the need to obtain access to the Pennsylvania Health Department's carefully guarded data.

Earlier in the summer I had been invited to address a public meeting in Harrisburg on the likely health effects of the accident at Three Mile Island. At the end of the meeting, someone introduced himself to me, and asked me whether he could be of any help. His name was Warren L. Prelesnik, and he told me that he was deeply concerned, since he had just moved to the area with his family shortly before the accident, and that he was working in the Harrisburg Hospital as executive vice-president in charge of administration.

I asked him whether it might be possible to obtain information on the monthly number of births and infant deaths together with their cause over the past few years, and he said that he would try to see what he could do.

More than a month later, when I had already given up hope of receiving any information, there arrived in the mail a letter with a list of the monthly infant deaths, fetal deaths, stillbirths, and live births in the Harrisburg Hospital for the previous two years.

At first there seemed to be no obvious change, if one looked at the total numbers of all types of fetal and infant deaths combined. But then I examined separately the category of newborn or neonatal infant deaths—those that were born alive but died within the first year, but mainly in the first few hours after birth. Here was the evidence I needed. In February, March, and April of 1979, there had only been 1 infant death per month. But for each of the two months of May and June, there were 4. Effectively, since the number of births had not only remained nearly the same but had actually declined slightly, this was more than a fourfold increase in the mortality rate, or of the right magnitude required to fit the observed 50 percent rise in the more distant area of upstate New York.

From an average of 5.7 per 1000 live births in the three months of February, March, and April—before the releases could have had an appreciable effect—the newborn mortality rate had risen to 24.1 for May and 26.0 for June, an unprecedented summer peak that did not occur the previous year. In fact, for May and June of 1978, there had been a total of only 3 infant deaths, while for the same period in 1979 after the accident, there had been 8.

As some of my colleagues with whom I discussed these findings agreed, by themselves the Harrisburg Hospital numbers were of course small, and only marginally significant, representing only about one-third of all the births and deaths in Harrisburg. But taken together with the vastly more significant and independent numbers for all of Pennsylvania, upstate New York, New York City, New Jersey, Maryland, and Ohio, there was now a much greater degree of certainty: It would have been much too much of a coincidence—perhaps less than one in a million—for all these different numbers to show the pattern they did.

At this very moment, there arrived an invitation to address an international meeting of engineers and scientists in Israel. The meeting was devoted to studying the environmental problems of industrialization of Third World nations. It was to be held in December, and I was asked if I would be willing to address the meeting on the environmental health problems connected with nuclear energy.

I could hardly believe that such an opportunity to bring these facts into the open should come at precisely the time when the data from the Harrisburg Hospital had convinced me more strongly than ever of the great danger of nuclear reactors. I decided to accept the invitation and to devote myself in the remaining few months to preparing a detailed paper that would begin with a review of the previous evidence on the effects of low-level radiation from fallout and normal nuclear plant releases and end with the evidence for the rise of infant deaths after Three Mile Island.

One of the remaining important questions that had to be checked, however, was the time and cause of death. Clearly, if the excess deaths were connected with the radioactive iodine released from the plant, then they should be associated with underweight births or immaturity, since damage to the fetal thyroid would slow down the normal rapid growth and development of the baby in the last few months before birth. The development of the lungs, which have to be ready to begin breathing at the moment of birth, is one of the most critical phases of late fetal development. Any developmental slowdown would be most life-threatening if it led to the inability of the tiny air sacs in the lungs to inflate and start supplying the blood with oxygen. Failure of the lungs to function properly would therefore lead to immediate symptoms of respiratory distress, and if efforts to treat the baby should not succeed, it would die in a matter of minutes, hours, or days of respiratory insufficiency or hyaline membrane disease.

Thus, one would not expect to find as large an increase in spontaneous miscarriages well before birth as newborn deaths within a short time after birth, since the lungs did not need to start functioning until the baby was born. Also, there should be no significant increase in gross congenital malformations a few months after the accident, since by the time the baby in the mother's womb had reached the sixth or seventh month of development, all the major organs had already fully developed. Thus, only some six to seven months after the accident would one expect some increase in serious physicial malformations, since these infants would have been exposed to radiation

in the first three months of development of critical-organ formation.

The data from the Harrisburg Hospital supported these
expectations. There was much less of an increase in the number
of spontaneous miscarriages and stillbirths than in the number
of newborn babies that died shortly after birth because of immaturity and respiratory distress, indicating the strong likelihood
that it was the effect of iodine 131 and other shorter-lived
iodines such as iodine 133 that had damaged the ability of
the thyroid to produce the necessary hormones needed for normal growth and development.

In fact, it was for this reason that I had publicly urged
widespread screening for hypothyroidism at the time of the
news conference in Harrisburg on the second day of the accident, the kind of simple test that could prevent permanent
mental retardation if detected and treated early. This test was
already being used routinely for every newborn baby born in
hospitals of a number of states in New England, the Northwest
Coast, and Pennsylvania, but not yet in New York or Maryland.
There would have to be a rise in the incidence of this condition
if my past findings on the increase in underweight births and
subtle forms of mental retardation during the period of heavy
nuclear-bomb testing were indeed related to the action of radioactive fallout. But not until many months later, long after the
Kemeny Commission hearing had been completed, would I
learn that a rise in hypothyroidism had already been discovered
by the Health Department of the State of Pennsylvania among
the newborn babies in areas where the invisible radioactive
gases from Three Mile Island had been carried by the winds.

19

The Present Danger

THE POTENTIAL EFFECT of the radioactive iodine on thyroid function and mental development was very much on my mind at that moment, since in early September I had presented our findings on the relation between fallout from bomb-tests and declines in the S.A.T. scores at the annual meeting of the American Psychological Association in New York. The likelihood that there would once again be widespread damage to the learning ability of children in areas reached by releases of radioactive fission gases, this time from peaceful nuclear plants either during normal operations or as a result of accidents such as the one at Three Mile Island, was in a way more disturbing than the evidence for rising infant deaths and later cancers.

The nation could survive if there were a few more infants that died shortly before or after birth. It could even survive if there were many more adults who would die of cancer or heart disease at age seventy rather than at eighty. But no nation could survive in the long run if it continuously damaged the mental ability of its newborn children, especially in an age where verbal and mathematical skills were increasingly essential to the functioning of a high-technology society. And since fewer children were being born, and the advances of modern medicine had increased greatly their chances of survival to adulthood even if they were physically and mentally handicapped, it would not take much more than a few generations for a nation with nuclear plants near its cities or sources of milk and water to destroy its health, its productivity, and thus its ability to compete with others who used less biologically damaging ways to meet their needs for energy.

Therefore, when I received an invitation to present my most recent findings at a meeting of the Connecticut Parent-Teachers Association in Hartford a few weeks later, I decided to accept. Hartford was not more than 40 miles northwest of the Millstone Nuclear plant, whose iodine 131 emission back in 1975 was officially listed as 10 curies by the NRC. This was almost as great as the amount admitted to have been released at Three Mile Island by the utility's own environmental consultants. For this reason, I decided to present my findings on the effects of Millstone as a way to estimate what the future health effects of the accident in Harrisburg were likely to be.

Strangely enough, it was through my concern about the possible effect of the October 1976 Chinese fallout discovered in southeastern Pennsylvania by the operators of a nuclear plant on the Susquehanna River not far from Three Mile Island that I first learned of the high releases from the Millstone reactor.

Apparently, as in the case of the Albany-Troy episode back in 1953, a heavy rainstorm brought down very large amounts of fallout from a nuclear cloud, setting off radiation alarms at the Peach Bottom Nuclear Power station near the Maryland border. That rainout had caused the evacuation of many of the workers from the plant. The EPA had failed to warn either the public, state health authorities, or the reactor's health physicists of the potentially high local fallout, hoping that it might not happen. Only when the plant supervisor got in touch with Thomas Gerusky at the Pennsylvania State Bureau of Radiation Control and checks were made at other locations such as the Three Mile Island plant did it become clear that the high iodine 131 levels were due to fallout, and not an accident at Peach Bottom.

When the iodine levels in the milk started to climb to a few hundred picocuries and no one had warned the public that pregnant women should not drink the milk, a colleague of mine at the University of Pittsburgh and I decided to hold a news conference to issue such a warning.

As it turned out, Gerusky decided not to order the cows to be placed on stored hay, even though some areas in Pennsylvania reached levels close to 500 picocuries per liter. Only in Massachusetts and briefly in Connecticut and New York did the health departments order dairy cattle to be switched to uncontaminated feed, and only in Massachusetts and Rhode Island, which obtained most of its milk from Massachusetts, did infant mortality continue its sharp decline in the following few months among all the New England states.

When a news story with my findings on the rises in infant mortality following this episode was published by the *Washington Post–Los Angeles Times* News Service in the summer of 1977, I received a phone call from a newspaper reporter in Connecticut, who asked me whether I had examined the possible effect of the Millstone plant releases on the pattern of infant mortality changes in New England. Someone had given him a copy of a recent annual environmental report for this plant, and he wondered whether I might be willing to look at it for him since he was unable to interpret its significance.

When the report arrived a few days later, I turned to the pages dealing with milk measurements. I could hardly believe my eyes. The control farms located in a direction where the wind rarely carried the gases from the stack showed levels of strontium 90 of only 5 to 7 picocuries per liter, similar to the rest of the East Coast. The concentrations in other nearby farms, however, reached values as high as 27 of these units, higher than those typical for Connecticut during the height of nuclear-bomb testing back in the early 1960s and similar to the highest concentrations measured by N.U.S. at Shippingport. For the people living within 10 to 20 miles of the plant, nuclear-bomb testing might just as well have never ended.

And when I looked at infant mortality in New England in preparation for a lecture at the University of Rhode Island, the familiar pattern I had seen at Dresden, Indian Point, and

Shippingport once again confirmed the seriousness of these levels of fallout in the milk. While throughout the 1950s and
1960s all the New England states had shown the same infant
mortality rate, following the onset of releases from Millstone
in 1970, Rhode Island, directly downwind, suddenly stopped
declining as rapidly as all the other states. By early 1976, before
the October fallout arrived from China, Rhode Island had
nearly twice the infant mortality rate of New Hampshire.

Shortly after I presented these findings at the University
of Rhode Island, I received a telephone call from State Representative John Anderson of the Connecticut legislature, asking
me whether I would be willing to undertake a more detailed
study of the possible health effects of Millstone and the nearby
Connecticut Yankee Reactor at Haddam Neck for the people
of Connecticut. I agreed on the condition that he would send
me the full environmental reports for the two plants for every
year of their operation, together with the detailed annual vital
statistics reports of the State of Connecticut.

A few weeks later a large box arrived containing the reports.
The story they revealed was a repetition of what had taken
place at Shippingport, except that this time the environmental
and health data were much more detailed and extended over
many years before and after the start of operation. Again, the
strontium 90 levels in the soil and milk increased as one approached each of the two plants. The levels were a few times
higher near the Millstone Plant, with its boiling-water reactor
(BWR), than near the Haddam Neck plant, with its pressurized-
water reactor (PWR), which was similar to Shippingport and
Three Mile Island.

This time, however, data was available for every year of
operation on a month-by-month basis, and it was possible to
see how in the first few years of operation, the strontium 90
levels were no different near the plants from those in the rest
of New England. But gradually, as the fallout from bomb testing
was washed into the rivers and the ocean by the rains, the
soil and milk levels declined all over New England, while they
stayed high or even rose for the farms within a 10- to 15-
mile radius of the plants.

On a number of occasions, when there was a particularly heavy fallout from a Chinese nuclear test, as in October of 1976, the records of the milk measurements showed the arrival of the fallout very clearly as a peak, particularly for the short-lived iodine 131 and strontium 89, and to a lesser degree for the long-lived cesium 137 and strontium 90. But what was even more disturbing were the even larger peaks of strontium 90 and cesium 137 in July and August of 1976, months before the bomb was detonated, not only in the local farms but as far downwind as Providence, Rhode Island.

Yet the summary in the front of the utility's environmental report for 1976 maintained, as it had every year, that the strontium 90 and cesium 137 in the milk was attributable to fallout from nuclear testing. It was sad to see that the once so hopeful nuclear industry now needed the continuation of nuclear-bomb tests to stay in operation.

To calculate the radiation doses to the bones of children, I used the high local excess values of strontium 90 in the milk along with the NRC's own calculational model given in NU-REG 1.109. The results were of the order of a few hundred millirems per year, many hundreds of times the value of less than 1 millirad arrived at by the utility when the strontium 90 was left out of the calculations, and far above the maximum of 25 millirems per year that was proposed by the EPA as the maximum permissible value from the nuclear fuel cycle.

Thus it was no surprise that the EPA as well as the NRC issued statements after my reports had been sent to State Representative Anderson and Congressman Christopher Dodd, in whose district the Millstone Plant was located, which claimed that the high strontium 90 and cesium 137 levels in the milk near this plant were due to fallout and could not be attributed to releases from the plant. The EPA and NRC never even attempted to explain why the levels of these radioactive substances should increase as one approached the stack from every direction.

Instead, these government agencies, on whom the public depended for the protection of its health and safety, tried to mislead the public. They claimed that there was little strontium

89 present along with the strontium 90, as is always the case when fresh fission products escape into the environment, and that therefore the strontium 90 could not be due to plant releases.

But what the nonspecialist could not have known is that strontium 89 has a very short half-life of only 50 days compared with 30 years for strontium 90. While the long-lived strontium 90 continues to build up in the soil around the plant, the strontium 89 rapidly decays away. Thus, when the cows return to pasture in the spring and summer, the milk shows predominantly the accumulated strontium 90, and very little of the short-lived strontium 89.

In fact, it is just as in the case of a coal-burning plant, where both steam and dust are emitted from the stack. Clearly, one would not expect to see the surrounding area covered with water, which evaporates rapidly just as short-lived isotopes disappear. Instead, one would expect to find a high level of ashes accumulating, decreasing with the distance away in every direction, just like the long-lived strontium 90 particles in the soil and milk around a nuclear plant.

But the nuclear scientists and engineers in these agencies, taking advantage of the widespread lack of scientific knowledge among the general public, its representatives, and even the heads of their own bureaucratic organizations, acted to protect the national interest as they saw it. Thus, they used their expertise to mislead the public, firmly believing that the need for energy independence or the willingness to use nuclear weapons far outweighed any conceivable small impact on human health.

Having the detailed figures on the officially announced releases as well as the uncontested measurements of radioactivity in the milk around Millstone over the years, I could compare the releases directly with those from Three Mile Island. Over a period of five years, Millstone had released half as many total curies of radioactive gases of all types into the atmosphere as Three Mile Island did in five days, including roughly comparable amounts of iodine 131. According to the health statistics, infant mortality in Rhode Island, some 20 to 40 miles away, was twice as great as for the most distant states after Millstone had operated for five years. Therefore, in my first approximation, there would have to be at least a 50 to 100 percent rise in infant mortality and childhood cancers in the Harrisburg area, which would be followed in the decades to come by cancer rises among the older population, perhaps leading to as many as 4,000 to 8,000 extra cancer deaths in the next few decades.

There was no need to extrapolate from very high doses to very low doses, since the amounts released in both cases were comparable. Both for the Millstone and Three Mile Island releases, the doses were in the range of tens to hundreds of millirems per year, and they were due to comparable types of radioactive elements created in the course of nuclear reactor operations.

But precisely because the releases from Three Mile Island were not so very different in magnitude from what the NRC and EPA had set as permissible for normal nuclear reactor releases in the course of a year, it was clear to me that enormous efforts would have to be made both by the government health agencies and the nuclear industry to keep knowledge of the likely health effects of the accident from reaching the public or their elected representatives in Congress. And this is precisely what happened in the weeks after my talk in Hartford, when the long-awaited report of the Kemeny Commission was being prepared in its final form.

I had been approached by ABC to appear on the show *Good Morning America* to present my findings, which were apparently in sharp contrast to the conclusion of the Kemeny report, a draft of which was read to me by the producer. According to this draft, which had a discussion of potential health effects that was confined to only a couple of pages, the only effects were psychological, with no detectable increases expected on infant mortality or cancer rates. In effect, the Kemeny Commission had accepted the optimistic report by the NRC, the EPA, and HEW a few days after the accident.

Apparently no efforts had been made to look at the actual statistics on infant mortality and miscarriages that had shown significant rises as early as May and June, four to five months before the final draft was being prepared in September and October. Yet, if the commissioners had wanted to, they could easily have obtained the same data I had been able to find in the records of local hospitals and the reports of the U.S. Center for Health Statistics for every state in the United States. If there really had been no increase in stillbirths and infant deaths, this would surely have been the best way to reassure the people of Harrisburg and the rest of the world living near nuclear reactors, once and for all ending the concern about nuclear power, silencing the critics, and freeing the industry from the uncertainty that was leading to its rapid decline in the wake of Three Mile Island.

But this was clearly not the course chosen. The actual data would have shown an increase in mortality rates near the plant during the summer months, while they declined in nearby areas not reached by the plume so carefully recorded in the utility's own internal reports available to the Kemeny Commission. Such a pattern would have been as difficult to explain away as the peaks of strontium 90 infant mortality and cancer increases around Shippingport and Millstone in the past, hardly reassuring for a public that had by now learned to distrust deeply the public statement of utility officials and government scientists whenever it came to the health effects of low-level radiation from bomb fallout or nuclear facilities.

Not being able to allow the truth to emerge, the government and the industry resorted once again to the familiar tactics of suppression and attempts to discredit the critics, as I would learn in the days following the official release of the Kemeny Report in early November.

I was supposed to appear on *Good Morning America* the day after the Kemeny Commission report was published. All arrangements had been made when I received a phone call from the producer saying that the format of the show would have to be changed, that they would need to find someone who would represent the industry and government point of view to debate me, and that this would mean a day's delay in my appearence. The following morning, I received another call from the producer, who said that they had found someone who would represent the other side, and that the program was now scheduled for 8:15 A.M. the next day. My tickets had been paid for, the hotel room in New York reserved, and a limousine ordered to pick me up and take me to the studio.

But the opportunity to present the other side of the story to a nationwide audience in answer to the bland assurance of the Kemeny Commission broadcast the day before never came. Just a few hours before I was scheduled to leave for New York, a call came from ABC saying that there was a last-minute change in the schedule, and that they had to cancel my appearance. I remembered the enormous pressures that had been exerted by the Atomic Energy Commission on the producers of the NBC *Today* show back in 1969 when I was scheduled to appear to talk about the effects of bomb fallout on infant mortality. But this time, it seemed likely to me that the pressure came from a commercial nuclear industry fighting for its life, and apparently these forces were too powerful even for a large television network such as ABC.

A news conference had been arranged by a local citizens' group in Harrisburg for noon, following my scheduled appearance on *Good Morning America,* and so instead of flying to New York, I took the plane to Harrisburg early the next day. It was the same flight I had taken the morning of the first news conference, when the radioactive gases were causing my survey meter to give me the warning of the large gas releases that the industry did not want to become known.

The news conference took place in the same small room of the Friends' Meeting House where the first one had been called on the second day of the accident. Dr. Chauncey Kepford, who had been one of the first scientists in the area to warn of the danger of the Three Mile Island plant, long before it went into operation, summarized his findings that the radiation doses were much larger than had been calculated from the simplified mathematical models used by the NRC and adopted by the Kemeny Commission. Because of his efforts to warn the local group of concerned citizens to prepare their case against the plant, he had been fired by Pennsylvania State University, something that he had been able to prove in court when he had sued the university for damages. Now he had nothing more to lose, and so he was able to provide independent evidence that the health effects of the accident would be much greater than the public had been led to believe.

I then outlined the substance of my findings based on the state-by-state data in the U.S. Monthly Vital Statistics. The data showed sharp rises in infant mortality in Pennsylvania and the area of New York State outside New York City, while the rates continued to decline in the areas of Philadelphia and New York City, where the radioactive plume had not been carried by the winds. I also cited the evidence of higher infant mortality rates in the Harrisburg and Holy Spirit Hospitals (in the Harrisburg area) for the months following the accident as compared with the same period a year before.

To this I added the latest findings that the rate of infant deaths had also gone up as far away as the Pittsburgh area, toward which the radioactive gases had drifted in the early period of high releases as recorded in the records of the largest hospital (namely the Magee Women's Hospital, associated with the University of Pittsburgh). According to the records of the hospital, which accounted for about half the births in Allegheny County, the number of deaths for the three-month period of May, June, and July had gone up 93 percent—from 27 in 1978 to 52 in 1979—while the births had remained essentially constant, rising only 2 percent—from 2166 to 2221. Furthermore, detailed examination of the causes of death revealed that the excess was due to an unexplained increase in prematurity, underweight births, and respiratory distress of the type found in the Harrisburg Hospital. There had not been any epidemic of other diseases or problems associated with the delivery process.

I concluded by saying that the evidence was therefore very strong that in the first few months following the accident, a few hundred excess deaths above normal expectations took place in Pennsylvania, contrary to the claims of the industry and the Kemeny Commission that there would be no detectable additional cases of cancer, developmental abnormalities, or genetic ill-health as a consequence of the accident at Three Mile Island.

There were a few questions after I handed out copies of

the U.S. Monthly Vital Statistics tables I had used, together with other tables and graphs summarizing the findings. The executive vice-president of the Harrisburg Hospital, Warren Prelesnik, who had given me the figures for his hospital, was present in the event that a reporter might wish to confirm the numbers, but no one inquired further, and the news conference broke up.

Television cameras representing the major networks had been present; some of the network reporters interviewed me separately immediately following the news conference. But neither that evening nor the next day was there any mention of these disturbing findings either on the local news in Pittsburgh or on any of the national television news programs. There were a few very brief local radio news items, but not a word of the news conference appeared in any Pittsburgh or Philadelphia papers.

It was as if an iron curtain had descended around the Harrisburg area, sealing off the people of the rest of the United States and the world from the news that would have warned them of a totally unexpected severe effect of low-level fallout. But neither the nuclear industry, the military, nor the state and federal governments committed to nuclear power wanted them to know. What so many people had feared would happen in a society committed to nuclear power had in fact taken place. The most important of all our civil rights, the freedom to learn of matters affecting our lives and those of our children through a free press, was being secretly subverted by an enormously powerful nuclear industry and a military establishment that had spawned and nurtured it, all in the interest of national security.

Since in our society there are so many independent magazines, newspapers, radio stations, and news services, unlike in a monolithic society such as the Soviet Union, there is no way to insure absolutely that a determined "dissident" scientist armed with publicly available government data can be prevented from having his message eventually reach the people. Therefore, the best way to prevent wide dissemination of undesirable information is to destroy the credibility of any individual seeking to reach the public and the scientific community at large. In this way, the message would either not be transmitted by wary news media or it would not be believed, especially if it was not reported in sufficient detail. This was, in fact, the tactic that was adopted following the news conference in Harrisburg. In the course of a detailed investigation for a story published in the June 1980 issue of the Canadian magazine *Harrowsmith,* one of its associate editors, Thomas Pawlick, a former investigative reporter for the *Detroit Free Press,* found out the following:

> First to attack Sternglass was *The Harrisburg Patriot* newspaper. A November 1979 article by Richard Roberts questioned Sternglass' figures on infant deaths in the city, charging that they did not "jibe with the hospital's statistics" as supplied by Harrisburg Hospital corporate relations officer Ernest McDowell. In a later unsigned editorial, the paper skirted the limits of libel, charging that the scientist was "inept at gathering statistics, or worse, he simply fabricated them to fit his conclusion." The editorial added: "For a scientist to present grossly inaccurate data is inexcusable. But to fit the method of analysis to a conclusion makes the scientist's motives suspect. Sternglass seemed principally concerned about his talk with the extent to which his appearance was documented by the media." There was no rise in infant deaths, concluded the *Patriot.*
>
> According to Harrisburg Hospital executive director Warren Prelesnik, who supplied the initial figures used by Sternglass, "no fabrication took place," and Sternglass' motives were far from suspect:
>
> "Dr. Sternglass used figures from a Hospital Utilization Project (HUP) computer read-out supplied by us," says Prelesnik. "The first set of figures we gave him, which he used in good faith, excluded, I believe, one or two cases (the discrepancy mentioned previously). This happened because we interpreted the term 'infant' differently than he did at first. We corrected this in a second set of figures I gave him on November 20, 1979. As for the figures quoted by Mr. McDowell, they came from a different source, that is, the hospital's Pediatric Mortality statistics."
>
> The Pediatric Mortality statistics include not only infant deaths, but those of older children—up to 11 years of age

in one case—as well as abortions and stillbirths. Averaged out, McDowell's figures would show little change after the accident. Only when infant deaths are isolated from the whole, as in Sternglass' report, does a post-accident rise show up.

The newspaper had undeservedly maligned Sternglass, whose claim of a rise in hospital statistics was correct. His figures for Magee Hospital in Pittsburgh were never questioned.

Nor did the article in the *Harrisburg Patriot* question or even refer in a single word to the highly significant numbers taken right out of the official U.S. Monthly Vital Statistics reports. But within a few weeks, nuclear industry spokesmen all over the world were quoting the *Harrisburg Patriot* editorial in attempts to discredit the paper which I delivered at the World Congress of Engineers and Architects in Tel Aviv, Israel, in December. This was especially the case in Sweden, where a great public debate on nuclear energy was in progress in connection with a referendum scheduled for early 1980.

But in the battle to restore my credibility, I received unexpected support from someone who had been on the inside of the Pennsylvania Department of Health, Dr. Gordon MacLeod. In fact, Dr. MacLeod headed the department during the period of the accident, having been confirmed in his position only twelve days before it occurred. He had been forced to resign shortly before the Kemeny Commission report was made public, and had returned to the University of Pittsburgh, where he headed the Department of Health Care Administration in the School of Public Health. As Pawlick described it in his *Harrowsmith* article:

In an interview with a reporter for *The Washington Post* [published February 2], MacLeod revealed that 13 babies (later corrected to 14) in three Pennsylvania counties in the path of the radioactive plumes had been born with hypothyroidism—ten more than would normally have been expected to occur. This initial figure was later expanded to include a total of 27 post-accident hypothyroidism cases throughout the whole state.

This disclosure prompted the state Department of Health to release its own figures, which confirmed that a higher than normal number of cases of hypothyroidism had been noted in the county immediately downwind of TMI.

On March 30, 1980, Dr. MacLeod went further. In a controversial speech delivered at Pittsburgh's First Unitarian Church, he stated that "recent data collected by the Pennsylvania Health Department show an increase in infant mortality within 10 miles of Three Mile Island when compared with the same population in the same time period for the preceding two years." He pointedly noted that this information had not been made public by the health department.

The exact figures for the population within 10 miles were 20 infant deaths in 1977, 14 in 1978, and a jump to 31 after the accident in 1979 for the six-month period April through September, while the number of births remained essentially constant. This meant that there had been a doubling in the infant mortality rate.

But of even greater importance, despite the relatively small size of the numbers, was the fact that for the area still closer to the reactor, the zone within a 5-mile radius, the rate had increased even more. In 1977 there had been only 3 infant deaths in this zone, declining to only 1 death in 1978. But in the six months after the release of the radioactive gases, the number rose sharply to 7 deaths, in close agreement with the sevenfold rise in the updated newborn death rate at the Harrisburg Hospital. Pawlick told the story this way:

> The Department of Health, its public credibility at stake, was forced to issue a news release April 2, in which it admitted that the rate of infant deaths per thousand live births "within ten miles of Three Mile Island" from April 1 through September 30, 1979 (after the accident) was 15.7, compared to a lower death rate of 13.3 per thousand for the entire state. The release, however, confused the issue by not comparing the April–September 1979 death rate with the *same period in 1978*. Instead, it recounted the figures for the period October 1978 through March 1979 (before the accident)—figures for the *winter* months, which are nor-

mally higher than summer months anyway. The rate for this rather irrelevant period was 17.2 per thousand.

Suddenly, despite all the efforts of the nuclear industry, the NRC, the EPA, the Pennsylvania Department of Health, and the *Harrisburg Patriot,* my findings had been substantiated by the most credible of all sources, the Secretary of Health of the Commonwealth of Pennsylvania at the time of the accident. Moreover, the Office of Vital Statistics of the Pennsylvania Department of Health had reluctantly confirmed his figures.

But the battle over the statistics at Three Mile Island was far from over; bigger guns would have to be brought into action. The very next day, *The New York Times* carried a story on these death rates headed: "No Big Change Found in Infant Death Rate Near Three Mile Island." It was a special interview with Tokuhata in which he claimed that studies just completed by his department found "no significant changes in these rates before and after the Three Mile Island Accident." As Pawlick pointed out, however:

> The Department of Health news release and the *New York Times* story both neglected to cite the figures for people living within a 5-mile radius of the nuclear plant that failed, as well as to cite the figures for the same months in earlier years. In an April 7 letter to the *Times'* editor, Dr. MacLeod made up for this deficiency. He did not reveal his sources, but it is supposed they were former colleagues within the Department of Health. (Indeed, a secretary of Dr. George Tokuhata, director of the department's Bureau of Health Research, admitted that the department's February 2 and April 2 news releases cited figures "that weren't originally intended for the public or the press, but the material was leaked. Somebody leaked the figures and we had to confirm them.")
>
> In his letter, MacLeod revealed that the infant death rates for those living within both a 10-mile and 5-mile radius of the stricken reactor had, indeed, risen sharply after the accident when compared to earlier years. In 1977, between April and September, the death rate was 6.7 per thousand within 5 miles of the plant and 10.5 per thousand within

10 miles. In 1978, the rates fell to 2.3 per thousand and 7.2 per thousand, respectively. But in 1979, after the accident, they jumped to 16.2 per thousand in a 5-mile radius of the plant and 15.7 within a 10-mile radius. MacLeod's figures substantially confirmed what Sternglass had been saying right along, that babies died in much higher numbers after the accident than they had been dying before it.

There was only one problem: *The New York Times* had refused to publish Dr. MacLeod's letter, which, with its crucial numbers, would have restored my public credibility.

Instead, the *Times* published a long article on the front page of its "Science" section by its reporter Jane Brody, on April 15, in which federal and state government spokesmen such as Dr. George Tokuhata tried to discredit both Dr. MacLeod's findings on hypothyroidism and my results on the rises in infant mortality.

What was particularly disturbing was the fact that neither Dr. MacLeod nor I was given an opportunity to reply to the statements of the spokesmen for the State of Pennsylvania and the U.S. Center for Disease Control in Atlanta cited in the *Times* article. MacLeod had received no call at all from Jane Brody, as I learned later, and the only questions I was asked had to do with the charge by Arthur Tamplin, the man who had been asked by the AEC to criticize my findings on infant mortality and bomb fallout back in 1969, that my studies were "incomplete."

Only three days later, still another attack was launched by the *Times,* this time on the editorial page. As Pawlick noted, the editorial was almost as insulting as the earlier editorial in the *Patriot,* accusing MacLeod of irresponsibility and me of "mishandling data," branding both of us as "Nuclear Fabulists" in its headline.

Thereupon, MacLeod sent another letter to the *Times* dated April 22, this time with a note to the editorial page editor, Max Frankel, asking whether *The New York Times* made it a policy to pillory individuals on its editorial page without giving them an opportunity to reply.

This time, MacLeod was told to call the science editor of the *Times,* William Stockton, who indicated that he had some problems with the letter, among which was MacLeod's listing of the actual numbers. Stockton also wanted to know why MacLeod did not disassociate his position on nuclear energy from mine, since he apparently was not opposed to nuclear power, something that MacLeod refused to do.

After further discussion, Stockton indicated that he would recommend publication of the letter only if MacLeod were willing to revise it. One alternative was to leave out the actual numbers that were subject to misuse, as well as a discussion of the sex ratio among the children born in the area cited by one of the federal government critics, which MacLeod had pointed out to have been completely erroneous. The letter as finally published by the *Times* on May 14 began as follows:

To the Editor:

Your April 18 editorial accusing me of telling nuclear scare stories and dealing recklessly with statistics is flawed by errors or omissions and ignorance of the facts.

The editorial, based on Jane Brody's news story three days earlier, can only revive public distress over data handling by state and federal officials. Had Jane Brody or your editorialist interviewed me, I could have told them immediately that I am not opposed to the use of nuclear energy. And I could have repeated that it is premature to blame the clustering of thyroid defects and the increase in infant deaths on the accident at Three Mile Island; but it cannot be ruled out yet, as federal and state officials have tried to do.

More than a year after the nuclear accident I released raw infant-mortality statistics which were then six months old. My announcement prompted the state to release infant deaths per thousand live births within 72 hours. Although both statistical measures are at best crude monitors of infant deaths, they should have been made public months before. After all, public health data belong to the public.

After pointing out that the data published so far did not include the numbers for the babies exposed in the first three months of development, when the embryo is especially sensitive to radiation, he continued as follows:

Yet to be explained is why 5- and 10-mile infant deaths around Three Mile Island during the six months following the accident climbed sharply, compared with the same period in previous years. In fact, the increases in the 1979 infant death rates over 1978 were statistically highly significant. Such significant increases in infant death rates following a nuclear reactor accident warrant complete candor and disclosure, not delay and denial. Had a decrease in infant deaths occurred, I trust it would have been widely publicized.

Here then was the crucial statement indicating that the numbers were found to be statistically highly significant by independent statisticians whom MacLeod had consulted, contrary to the claims of both Tokuhata and the scientists at the Center for Disease Control. But the actual numbers that would have convinced many skeptical scientists and laymen of the truth of what I had been saying over the years were left out. But so was any remark that could have been used to support the industry's attack on my credibility.

Turning next to the question of the significance of the increase in hypothyroidism, MacLeod went on as follows:

Despite the shortcomings in Brody's article, she contradicts your editorial undermining me for recklessly linking thyroid defects to radiation released from the crippled reactors. She stated, "Dr. MacLeod, however, did not attribute the cases to the accident."

I had expressed concern about a three months' delay by the Pennsylvania Health Department in announcing an unusual cluster of 12 times the expected number of hypothyroid cases in the county immediately downwind of Three Mile Island. My advice was accurately reported at the time as made only to encourage early detection of thyroid deficiency from any cause in unscreened newborns lest an untreated infant become a cretin.

The credibility of an official denial that radiation from the accident could have any effect whatsoever on fetal thyroid glands was clouded by an error made by an epidemiologist from the Center for Disease Control. In the news article, Dr. Greenberg mistakenly states that Pennsylvania had fewer newborns with thyroid defects than "all the areas in North America with screening programs."

That simply is not true. In the February 1979 issue of

Pediatrics, both Montana and Idaho had far lower rates of thyroid defects in newborns than Pennsylvania had following the nuclear accident. Also, Philadelphia and Pittsburgh have lower rates of thyroid defects in newborn than all of Pennsylvania.

Again, there were no exact numbers, but the article MacLeod had referred to, published by Dr. Stephen LaFranchi of the Oregon Health Science Center, showed that Montana had only one case of hypothyroidism for nearly 12,000 births. This was a much lower rate than for Pennsylvania in the nine months following the accident, when there were 27 cases in 119,000 births, or 1 case in 4400 live-born infants, a rate that was almost three times greater than in Montana. Yet for the nine months prior to the accident, Pennsylvania health authorities had discovered only 17 cases for about the same number of births. This was 37% lower, and thus closer to the rate of Montana, where there were no nuclear reactors at all.

Significantly, LaFranchi's paper also showed that Alaska, where according to the EPA's own data the heavier Chinese fallout had come down in recent years, showed a much higher incidence of hypothyroidism than Montana, namely a rate of 1 for every 3778 births, lending still more support to the hypothesis that low levels of iodine 131 were responsible for the increase in Pennsylvania following the accident. Furthermore, there was direct evidence of much higher thyroid doses than had been measured by the dosimeters from studies of field mice or voles examined by a group of independent scientists from nearly Millersville State College, preliminary reports of which had already become available. The iodine 131 found in the thyroids of these animals were in fact comparable to those found by the Utah scientists following the "Baneberry" accident back in 1970.

MacLeod concluded his letter as follows:

I am especially concerned about those citizens in central Pennsylvania who have lost confidence in the credibility of official statements since the accident over a year ago.

MacLeod had drawn attention to the real problem: the fact that it was the credibility of public officials entrusted with the health and safety of the people that had been destroyed.

When Pawlick called Dr. Frank Greenberg at the Center for Disease Control to obtain his reaction, he was unable to explain his error. Asked why such a misstatement had been made, he at first said, "Well, we can't really compare one state against the other. . . ." According to Pawlick's story, he then put him on hold. His secretary subsequently came on the phone to report that "Dr. Greenberg has been called away on an emergency." She added that it would not be worthwhile to contact him later.

Continuing the report on his investigation, Pawlick then turned to the way that Dr. Gary Stein of the CDC in Atlanta and Pennsylvania's Dr. George Tokuhata attacked the state's own figures in the *Times* article, using them as a sort of statistical "straw man" while ignoring much of the data reported by MacLeod and me.

Among the claims made by Tokuhata was that the Harrisburg infant mortality figures could not be used because the heavy black population had an unusually high infant mortality. That was exactly the kind of argument he had used in trying to explain away the high infant mortality rate in Aliquippa near the Shippingport plant seven years earlier, and it was equally misleading to the present situation. In both cases, it was the *change* in infant mortality that mattered before and after the releases, not its absolute value. It was clearly absurd to explain a doubling of the infant death rate by a sudden doubling in the black population or in the number of black babies born. In Aliquippa, the infant death rate had declined sharply within a year after Shippingport had been shut down, and there was no exodus of black people from the area.

Tokuhata also charged that my analysis had been based on "the wrong number" for the July 1979 infant mortality rate, which had been listed incorrectly in the U.S. Vital Statistics as 271 when it should have been 185. He claimed that this was an error made by the Center for Health Statistics in Washington due to a mix-up of fetal with infant deaths, which he had discovered after reading my Three Mile Island paper, as his secretary had told me when she had called me on March 14 a few weeks before the first *Times* story appeared on April 3.

Again, the nuclear industry, in its attack on my findings at Three Mile Island, learned of this "error" amazingly quickly. Only three days after the phone call from Tokuhata's office advising me of the mistake in the official statistics that conveniently reduced the number of deaths for July to just above the U.S. rate, the "error" was cited by a young Westinghouse engineer in a radio debate with me in Pittsburgh, well before any public announcement or article had reached the Pittsburgh news media.

As it turned out, even this adjustment of the published vital statistics did not alter the overall pattern of unprecedentedly high infant mortality rates for the entire period of May through December 1979 in Pennsylvania and in New York State outside New York City. The figures in the U.S. Monthly Vital Statistics for Pennsylvania told the story: Whereas shortly after the winter peak, or in February and March, Pennsylvania had an infant mortality rate 15 to 26 percent below that of the United States, for every month thereafter it exceeded the U.S. rate, even after the alleged mistake in July was corrected. As explained to me in a letter sent by Tokuhata early in April of 1980, the erroneously high figure for July was compensated for by reducing the number listed for August by 86 deaths, thus keeping the running total from January to August correct. Since the monthly figures are only regarded as provisional in any case, this method of correction is used rather than one in which each monthly figure is revised. This reduced the listed figure for August to 119, however, when in fact the August figure would have been 205 if no mistake had occurred in July. The highest excess for Pennsylvania had therefore actually occurred in August, when the Pennsylvania rate exceeded that for the U.S. by 34 percent. But the total number of infant deaths for July *and* August combined was unchanged at 390.

Thus, contrary to Tokuhata's claim in Brody's article that when the correct number was substituted "there was no increase in infant mortality last summer," there was in fact a large excess over the winter quarter relative to the rate for the U.S. as a whole. This can be seen more easily in the table below:

Pennsylvania and United States Infant Mortality (0–1 Year
old at death) 1979, With corrections in July and August*
(Data from the U.S. Monthly Vital Statistics)

	Pa. Deaths	Pa. Births	Pa. Rate Per 1000 Births	U.S. Rate Per 1000 Births	Excess of U.S.	Average Excess of Pa. Per Quarter
Jan.	216	13,112	16.5	13.8	+19%	
Feb.	147	11,892	12.4	14.6	−15%	−8%
March	141	13,589	10.4	14.1	−26%	
April	166	12,520	13.3	13.2	+ 1%	
May	198	13,201	15.0	12.6	+19%	+8%
June	163	12,293	13.3	12.9	+ 3%	
July	185*	14,680	12.6	12.5	+ 1%	
August	205*	13,918	14.7	11.0	+34%	+14%
Sept.	199	14,275	13.9	12.9	+ 8%	
Oct.	208	13,161	15.8	12.9	+22%	
Nov.	182	13,271	13.7	13.1	+ 5%	+13%
Dec.	175	11,871	14.7	13.2	+11%	

* The figures for July and August actually listed in the U.S. Monthly Vital
Statistics for these months are 271 and 119 respectively, which add up to
the same total of 390 as the corrected numbers.

If the figures for the State of Pennsylvania had followed
their decade-long pattern of infant mortality rates 5 to 10 percent below that of the United States in the summer of 1979,
as they did in the three months before the accident (−8 percent),
there would have been some 268 fewer infant deaths than the
1413 that actually occurred.

Moreover, when I added the figures for New York State
outside New York City where the August excess over the low
point in March was a startling 69 percent—namely 20.1 versus
11.9 per 1000 births—the total deaths above normal expectations for Pennsylvania and New York combined rose by 353
to the much larger figure of 621 for the accident at Three
Mile Island. And no one had as yet suggested any errors in
the published figures for New York State.

So great were the excess mortality rates for the rural and

small-town areas of upstate New York that I decided to try to check them with yet another set of data, namely figures reported weekly by the larger cities directly to the CDC in Atlanta and published in the Morbidity and Mortality Weekly Reports.

What I discovered there confirmed the findings based on both hospital records and the Monthly U.S. Vital Statistics. In city after city in the path of the invisible gas clouds drifting west and north, infant mortality in July 1979 had risen sharply over the same month in the year before the accident.

For Pittsburgh, it was a jump from 5 to 17 deaths, a rise of 240 percent. For Syracuse, the deaths rose from 10 to 18, or by 80 percent. And for Albany, the change was from 6 to 10, an increase of 67 percent.

But when I examined the figures for areas such as Philadelphia and New York City, not reached by the most intense clouds of radioactive gases, instead of rising, infant mortality declined, as it has been doing since the mid-1960s, when the fallout from the U.S. and U.S.S.R. atmospheric bomb tests began to decline. For Philadelphia, the number of infant deaths in July declined from 61 in 1975 to 50 in 1979. For New York City the decline was from 136 to 127.

Thus, still another set of official data confirmed my original findings as well as the data MacLeod had forced into the open, and I knew that Tokuhata's attempt to mislead the public in the pages of the Times would eventually backfire, still further increasing the tragic mistrust of the public for its institutions.

As Pawlick's Harrowsmith story put it:

The statistical battle still drags on, but more than numbers are involved in the controversy. The public's right to know is also at stake, as MacLeod insists. So is the future safety of Pennsylvania residents, as radioactive gases such as krypton 85 are vented into the outside air during the TMI cleanup process. (In an April 16 letter to the NRC, the Roswell Park Memorial Institute's Dr. Irving Bross called the krypton venting "a criminal action" and warned that it would "produce at least 50 excess infant deaths in the area.")

But in the account of the aftermath of Three Mile Island, there was one ray of hope: There were at least a few public officials not corrupted by the power of the nuclear industry, who had the courage to speak their minds even if it meant that they might lose their jobs, as Gordon MacLeod had. As Pawlick described him:

> Dr. MacLeod, sitting in a squeaky swivel chair in his linoleum-floored office at the University of Pittsburgh, is the very picture of academic reserve. Looking like he has been sent from Central Casting, he embodies arched-eye-browed, thin-lipped scientific respectability. MacLeod takes great pains to emphasize that he is not by any means anti-nuclear. He is simply pro-truth, and by now totally convinced that that was why he was fired.
>
> "The (Pennsylvania) health department is being exceedingly restrictive with regard to the release of data," he says, his chair emitting a high-pitched squeak as he rocks gently, meditatively from side to side, weighing his words. "There is also a timing problem because of the anxiety among the population. In the case of the hypothyrodism they knew about the data from October of 1979 and it wasn't until three months later that the public and the medical profession were alerted.
>
> "As for infant mortality, somebody [in the health department] had a hypothesis that the infant death rate would be affected within a 5- and 10-mile radius by the accident, and studies were launched and they are collecting data. If we have these kinds of figures for the six months after the accident [namely, those detailed in the April 2 health department press release], why haven't we gotten the preliminary data for the nine months yet? Only those children born later than September 1979 would have been exposed to radiation during the first trimester of pregnancy, that is, the time at which birth defects can occur. These later figures may be even higher."
>
> According to Dr. Tokuhata, the results of a study of "pregnancy outcomes" for women who conceived between March 1979 and March 1980 will not be released before 1982. (Tokuhata also described several other post-accident studies in progress, including studies of the psychological

effects of the accident on local residents. Although psychological data will doubtless yield much valuable information, their chief use to date seems to be to provide ammunition for industry defenders attempting to give the impression that any claims of health damage are really all in people's minds.)

"When I was called by former colleagues in the department and told these data, I asked if they were going to release them [the 5- and 10-mile radius data], and they said no, and I told them I was profoundly dismayed because the population is waiting to find out one way or the other," MacLeod continues. Any unusual patterns found in such studies "warrant complete candor and disclosure, not delay and denial," he insists, and delaying the release of data until 1982 could seriously hurt the "credibility of health professionals."

Ironically, it may have been his own attempt at candor and disclosure that cost him his cabinet post. "The Governor asked for my resignation on October 9, 1979," he recalls, noting that "the sequence of events" preceeding it was briefly as follows:

On October 4, copies of an environmental report prepared by the Governor's Commission on Three Mile Island and submitted later to the President's Commission on the accident were distributed at a meeting of the Governor's Commission members. The report quoted a Pennsylvania Department of Environmental Resource employee, whose testimony MacLeod found riddled with "misstatements and misrepresentations." On the advice of fellow state commissioners, he decided to point them out to the President's Commission and to the head of the Department of Environmental Resources, Clifford Jones.

As an example of some of the points the DER report left out, MacLeod notes that: "The maximum airborne iodine concentration occurred in mid-April, in connection with replacement of the auxiliary building charcoal filters and probably a far higher release of radio iodine occurred at that point, and we don't have the raw data for that [in the DER report]. It doesn't mention it."

MacLeod dictated a letter noting the report's shortcomings and sent it to the Presidential Commission, then he telephoned Jones. "I said, 'You've got a problem, Cliff, a man in your department who is misrepresenting the facts before the Presidential Commission, and this will embarrass the [Governor's] administration.' And he [Jones] got angry at me. That was Thursday. Then on Saturday I got a call from the governor's office to come in on Tuesday morning. He [the Governor] sort of uncomfortably asked me then for my resignation." Pressed for an explanation for the firing, MacLeod says Thornburgh "told me he had talked to Clifford Jones. I'm not sure whether it was a turf problem between the two departments, or what it was all about."

Asked about the firing, Thornburgh's press secretary, Paul Crithclow, said MacLeod was dismissed because "he had a great deal of trouble working with other cabinet officers and other administration officials." Without being asked, he volunteered the information that "his [MacLeod's] behavior during that period [that of the accident] was erratic," and insisted on reading extracts from the *Times'* "Nuclear Fabulists" editorial aloud over the phone.

MacLeod was replaced by H. Arnold Muller, a physician specializing in emergency medical care and the handling of battlefield casualties, with ten years of service in the military. Muller did not have any public-health background, but he did have a continuing financial connection to a project sponsored by the war college at a military base near Harrisburg.

At a meeting at the Pennsylvania Association of Hospital Auxiliaries at Hershey, Pennsylvania, early in May—just before the venting of the radioactive gases in the containment building had been approved by the Governor—he had said that fears stemming from the accident at Three Mile Island were unfounded. According to the story in the *Harrisburg Patriot* of May 7, Muller claimed that "there is nothing to indicate that there has been any illness whatsoever" as a result of the previous year's nuclear accident. Urging that the Three Mile Island accident should be put in proper context, he continued by saying, "Nobody died at TMI, nobody came close to dying." Referring to his experience in the case of automobile accident victims, he asked the women, "Where are all the crying people when a man dies on a street" as the result of a drunken driver? "There are none," he concluded.

Only a few days earlier, at a meeting on "The Roles of Local and State Health Departments in the Management of Radiological Emergencies" at the University of Pittsburgh School of Public Health, Muller had indicated, to the astonishment of everyone present, that the state would release only the health data that he could understand and approve.

On May 19, this department issued a news release prominently reported in *The New York Times* the following day under the headline "Fewer Infant Deaths Near Three Mile Island." In this article, the crucial figures for the infant mortality rate within a 5-mile radius around the plant—a rate that had increased sevenfold—were left out. Just like the misleading data that had been handed out to the *Harrisburg Patriot* reporter on the day of the news conference in November by the head of public relations for the Harrisburg Hospital, the data here had been lumped together with miscarriages and stillbirths under the category of "perinatal mortality." These rates hardly changed at all, especially when compared for the entire year from January through December of 1978 and 1979, because the effect of growth retardation on lung function does not show up until just after birth.

Instead, it was clear to me that the UPI story printed in the *Times* misled the public by comparing the low infant mortality rate for the high socio-economic, white suburban area within 10 miles of the reactor, which had been far below the rest of the state before the accident, with the figures for the state as a whole: "The figures show the infant death rate (deaths under one year) to be 11.5 per 1000 live births within a 10-mile radius of TMI. The statewide figure for the same period of time [all of 1979] was 13.3 infant deaths per live births." This was exactly the technique used by Tokuhata to hide the large rise in cancer rates around Shippingport compared with their previously low values relative to the state as a whole seven years earlier.

But if the public had been able to see the actual tables released by the Pennsylvania Department of Health broken down by quarters, they would have seen otherwise. In the crucial summer months after the accident, when the infants in their mothers' wombs that had functioning thyroid glands at the time of the accident were born, the infant mortality rate within the 10-mile zone had doubled, exactly as MacLeod had learned from deeply troubled health officials in Harrisburg.

Thus, for July, August, and September, when death rates are usually at their lowest, infant mortality rates within 10 miles had been 4.9 in 1978, and 12.8 after the accident in 1979, a rise of 160 percent. Nor was there any reference to the summer increases for Pennsylvania as a whole as reported in the U.S. Monthly Vital Statistics, far above the rates for the rest of the United States.

Yet the state's news release signed by Muller, as quoted in the UPI story, concluded as follows:

After careful study of all available information, we continue to find *no* evidence to date that radiation from the nuclear power plant resulted in increased number of fetal, neonatal or infant deaths.

Neither the UPI nor *The New York Times* had fulfilled their normal journalistic responsibility to the public to obtain comments from those who could have pointed out the misleading nature of the news release. In a matter of such great concern and importance for the future health and well-being of the children of Harrisburg and the entire world, was this too much to ask for?

Ironically, the willingness of *The New York Times* and the UPI to lend themselves to the attempt to cover up the full dimensions of the deaths at Three Mile Island was to be proven futile within a few months, as a result of the persistence of two television news reporters who became disturbed when they discovered a series of inconsistencies and anomalies in the tables of statistics released by the Pennsylvania Health Department in May of 1980. In going over the numbers for fetal deaths in the area within 10 miles of TMI for 1979, they noticed something that was mathematically impossible. They found that for the month of January, the average fetal death rate listed for the entire area was *smaller* than either the rate in the city of Harrisburg or the rate for the suburban area within the entire 10-mile zone taken separately.

Their suspicions aroused, they continued their detailed examination of the official figures and noticed that the rates of fetal deaths with and without therapeutic abortions for the suburban portion of the 10-mile zone were exactly identical month for month through all of 1979. This was very strange indeed, since it meant that there was not a single reported induced abortion for all of that year in the area closest to the reactor, where the concern of pregnant women was greatest.

Even stranger was the fact that for the city of Harrisburg, the table showed a difference between fetal deaths for the period April to September with and without induced abortions only for May. This meant that there was only a single induced abortion listed for the summer of 1979 in any part of the heavily populated 10-mile zone around Three Mile Island. Yet, another table in the released material showed 11 induced abortions in 1978 and 10 in 1977 for the same 10-mile area around the stricken plant during the same six-month period.

Deeply troubled by their findings, the two reporters went to see MacLeod about a possible explanation. He agreed that there was something wrong with the numbers as listed, and promised to check into the matter further, which he did by consulting a number of statisticians at the University and the State Health Department. None of the individuals he consulted could give an explanation, and one person who had access to the original data as kept in the State's computer said that the numbers were altered in the release.

Shortly thereafter, I happened to stop by at MacLeod's office, when he told me of these very disturbing facts. Apparently it had not occurred to either one of us that the data released might actually have been doctored in some manner when we first heard of the May 1980 news release issued by MacLeod's successor. It seemed incredible that someone might want to do something so glaring, yet when we examined the figures further, a whole series of gross inconsistencies emerged, all tending to reduce the number of deaths during the critical summer months when the U.S. Monthly Vital Statistics had shown the greatest rise of infant deaths both in Pennsylvania and upstate New York relative to the United States as a whole.

As to the sudden decline of therapeutic abortions in the

Harrisburg area, MacLeod had called some of his physician friends in Harrisburg and learned that they had performed just about the same number in 1979 as in 1978. And when I looked at a summary for the Harrisburg Hospital sent to me earlier by Prelesnik, there were a total of 10 therapeutic abortions listed for April through September in 1978 and 21 for 1979, totally at variance with a 90% drop to essentially none for 1979 listed for this period in the May release of the Health Department.

Remembering the mistake of an excess of 86 infant deaths in July that was supposedly made and which Tokuhata said had been corrected in the U.S. Monthly Vital Statistics for August, I decided to check what the number of infant deaths were for July and August in the May 1980 Health Department release. Knowing the number of births which had apparently not been in error, it was a simple matter to calculate the number of infant deaths from the infant mortality rates listed. For July, using the rate of 11.9 deaths per thousand births and the 14,680 live births, one obtained 175 infants that had died according to the May 1980 release. For August, the listed rate of 10.2 and the 13,918 births gave 142 infant deaths, for a combined total in July and August of 142 plus 175, or 317.

But this was 73 deaths less than the 390 given in the U.S. Monthly Vital Statistics for the two months after the alleged error had been corrected. The number of infant deaths listed had quietly been further reduced so as to get a still lower rate.

For the three summer months of July, August and September, the U.S. Monthly Vital Statistics gave 589 infant deaths after the August correction had been made. But the figures released by Muller in May 1980 gave a total of only 501 infant deaths, 88 fewer than had been reported to Washington in 1979. Altogether these two adjustments reduced the number of infant deaths by 174, compared with the figures originally reported to Washington in the summer of 1979. This resulted in a low infant mortality rate of only 11.7 per thousand births instead of 13.7 for the summer quarter, bringing it down to below the U.S. rate for the summer quarter of 12.1. Thus it would be possible to maintain Muller's claim in the official release that "after careful study of all available information we continue to find *no* evidence to date that radiation from the nuclear power plant resulted in an increased number of fetal, neonatal or infant deaths."

The damage done to the developing infants at Three Mile Island will not be as easily swept away as a single public-health official, more concerned about trying to protect human life and health than a powerful technology gone out of human control.

I knew only too well how often this had happened before without the knowledge of the public. I knew how the budgets of public-health agencies, such as those of New York State, had been cut in order to stop the publication of the detailed annual health statistics that would allow other conscientious officials or independent investigators to alert the public to the danger of emissions from newly built nuclear reactors or fallout from distant nuclear detonations. The fragmentary summaries of data that replaced the detailed reports beginning in 1970 were a very inadequate substitute. I also knew that the budget of the EPA had been cut by the Nixon administration to force an end to the publication of *Radiation Health Data and Reports* in 1974. That was the year after the nuclear industry and the agencies that promoted it had learned from the Shippingport hearings how the detailed monthly data on strontium 90 gathered by the states could be used to pinpoint the new sources of radioactivity in the milk. Used intelligently, such detailed data might lead to costly damage suits, just as in the case for the fallout from Nevada.

After those who were primarily concerned about public health had been forced out of the NRC and EPA, it was a simple step to end the previously required monitoring of strontium 90 by the nuclear plants, ostensibly as an economy measure. Interestingly, however, the end of monitoring came in 1979, the same year in which the permissible doses to critical organs from the nuclear fuel cycle were reduced by a factor of twenty. Those scientists who knew that strontium 90 gave the greatest dose per picocurie of all substances released by nuclear bombs or nuclear reactors would no longer be able to protect the public precisely because the most crucial data was no longer being collected. And those few who wanted to warn the public risked the destruction of their scientific reputation and careers.

As I explained to Pawlick at the end of our interview, one of the greatest unanticipated threats of low-level radiation to the human body comes from its action on normal, life-giving oxygen molecules, turning them into powerful toxic agents. Among the most important systems they attack are the immune defenses of the body, which detect and destroy not only foreign bodies such as viruses and bacteria, but also ordinary cells that have somehow gotten out of normal control. These are the so-called malignant cancer cells, which multiply rapidly until they become so numerous that they inhibit the normal functions of vital organs, a condition that eventually leads to the death of the organisms as a whole.

In this sense, there is a close analogy between the human body and a complex human society. They can both be destroyed by outside forces, or they can destroy themselves if they lose the ability to recognize "super-normal" individuals with an unusual ability to propagate their kind in an unchecked manner.

In our rapidly changing science-based society, it is the freedom to investigate and communicate important scientific or public health findings quickly and widely—no matter how disturbing or controversial—that is the key element in the protective system needed to alert a society to potentially dangerous developments before they become irreversibly destructive.

The rapid growth of a powerful military and commercial nuclear technology was largely unchecked by the normal protective processes of free communication and public discussion. As a result, the unique economic and political forces of the industrial, military, and scientific organizations to which the atom gave birth are like a malignant cancer in our society, unrecognized and unchecked while it developed under the cover of secrecy to its present enormous size. If we continue to allow our government, which brought this technology into being for purposes of national security, to continue in its efforts to aid and abet the suppression of the freedom of publication in this vital area, then the crucial early warning system that our society needs to survive will have been destroyed.

In the name of national security, our scientists and engineers have created Frankenstein's monster, capable of destroying life in this world. Ironically, in order to realize the dream of ending all wars and developing the peaceful atom that would atone for the horror of Hiroshima and make up to mankind for the threat of destruction that would forever hang over the world in the years to come, they needed to ally themselves with the military, political, and economic interests that alone could supply the enormous financial resources needed to realize their dream. Indeed, Eisenhower had tried to warn the nation of this danger at the end of his presidency.

The alliance of science and technology with the military and political forces is, of course, as old as civilization itself, since only through the fear of powerful enemies would the public provide the necessary funds to develop costly new technologies, all the way from better steel for swords to gigantic missile systems capable of pinpoint accuracy in delivering nuclear bombs to their targets.

But when the testing of nuclear weapons and the leakage from commercial reactors were found to have unanticipated serious biological effects on the population, it became necessary to secretly subvert the very freedom of publication and continued correction of errors on which the success of modern science and technology itself has been based.

In their understandable desire to see the blessings of the peaceful atom come about in their lifetime, and concerned not to endanger the sources of capital for the research and development essential for the advancement of science and technology required by modern society, those involved with the development, promotion, and regulation of nuclear technology and the protection of public health were too often willing to participate in the effort to hide the consequences of nuclear testing or normal and accidental releases from nuclear reactors, especially when the requirements of national security were cited to them in periods of international tension.

Ironically, the need to believe that peaceful applications of the atom were possible played into the hands of those in the military who wanted to use nuclear weapons in limited wars, since both required the assumption that low-level radiation from distant, worldwide fallout or from nuclear plants was essentially harmless. Thus, the most concerned and idealistic scientists who had worked on the bomb and who later dedicated themselves to the realization of the peaceful benefits of the atom, because they were willing to believe the harmlessness of very small amounts of radiation and the negligible magnitude of the doses from nuclear reactor operations, were in effect contributing to the increased likelihood of nuclear war.

Thus, the deeply felt hope for safe, clean, and economical nuclear power kindled by the nuclear scientists tragically aided the plans of leaders of the nuclear nations to find ways to use nuclear weapons in all types of military confrontations. Only the continuing denial of the seriousness of worldwide fallout would give credibility to these threats.

Only a few months before Three Mile Island, James Reston, writing in *The New York Times,* asked what "the present danger" facing our nation really was:

> Is it a military threat from the Soviet Union or an economic threat from some of our allies who are outworking and outproducing us?
>
> In short, is the threat external or internal? What worries the world about the United States today: that it is spending only 117.3 billion dollars this year on defense—the highest peacetime military budget in our history? Or that the United States is spending more of its economic and moral capital than ever before and losing confidence in itself and the confidence of the free world?

Reston went on to quote Lincoln from an address given in Springfield, Illinois, on January 27, 1837. Lincoln's words now take on a particularly strong relevance:

> At what point shall we Americans expect the approach of danger? By what means shall we fortify against it? Shall we expect some trans-Atlantic military giant to step the

ocean and crush us at a blow? Never! All the armies of Europe, Asia, and Africa combined, with all the treasure of the earth (our own excepted) in their military chest with a Bonaparte for a commander, could not by force take a drink from the Ohio or make track on the Blue Ridge in a trial of a thousand years.

And then came this most strangely prophetic passage:

At what point then is the approach of danger to be expected? I answer, if it ever reach us it must spring up amongst us: it cannot come from abroad. If destruction be our lot, we must ourselves be its author and finisher. As a nation of free men we must live through all time or die by suicide.

As Reston concluded, "it could be, of course, that Mr. Lincoln is out of date in this nuclear world, but at least his point is worth debating. The 'present danger' may be the failure to debate what it really is."

But when vital information is secretly kept from free people, they are no longer free, and there can be no meaningful debate of the most crucial problem facing our nation and the rest of the people of this world, namely whether we shall learn how to live through all time by finding a way to end the nuclear cancer threatening our nation, or die by nuclear suicide.

If we have any moral or ethical obligations at all as human beings, they surely include the obligation to insure the survival of our species and thus the opportunity for our children and their descendants to develop to the fullest the miraculous potential of the human mind. As the French philosopher-scientist Jean Rostand has phrased it so eleoquently on behalf of humanity as a whole, "The duty to survive gives us the right to know."

Bibliography

Arakawa, E. T. 1960. Radiation dosimetry in Hiroshima and Nagasaki atomic bomb survivors. *New England Journal of Medicine* 263:488–493.

Beierwaltes, W. H., et al. Aug. 27, 1960. Radioactive iodine concentration in the fetal human thyroid gland from fallout, *Journal of the American Medical Association* 173:1895.

Berke, H. L., and Deitch, D. April 1970. Pathological effects in the rat after repetitive exposure to europium 152–154. *Inhalation carcinogenesis,* AEC Symposium Series, vol. 18, pp. 429–431.

Bizzozero, J., Johnson, K., Ciocco, A., Hoshino, T., Toga, T., Toyoda, S., and Kawasaki, S. 1966. Radiation related leukemia in Hiroshima and Nagasaki. *New England Journal of Medicine* 274:1095–1101.

Boffey, Philip A., Gofman, John W., and Tamplin, Arthur R. Aug. 28, 1970. Harassment charges against AEC, Livermore. *Science* 169:838–843.

Cahill, D. F., and Yuile, C. L. May 1969. Some effects of tritiated water on mammalian fetal development. *Radiation biology of the fetal and juvenile mammal.* AEC Symposium Series, vol. 17, Proceedings of the 9th Hanford Biology Symposium, pp. 283–287.

Chase, Helen C., and Byrnes, Mary E. October 1970. Trends in prematurity in the United States. *American Journal of Public Health* 60:1967–1983.

Clark, Herbert M. May 7, 1954. The occurrence of an unusually high-level radioactive rainout in the area of Troy, N.Y. *Science* 119:619–622.

———— et al. November 1954. Measurement of radioactive fallout in reservoirs. *Journal of the American Water Works Association* 46:1101–1111.

Conard, R. A., Rall, J. E., and Shitow, W. W. 1965. Growth status of children exposed to fallout radiation on the Marshall Islands, *Pediatrics* 36:721.

————, ————, and ————. 1966. Thyroid nodules as a late sequel of radioactive fallout, *New England Journal of Medicine* 274:1392.

DeGroot, Morris H. 1971. Statistical studies of the effect of low-level radiation from nuclear reactors on human health. *Proceedings of the 6th Berkeley Symposium on Mathematical Statistics and Probability,* Berkeley, Calif., July 19–22, 1971, ed. J. Neyman. Berkeley: University of California Press.

Dyson, Freeman J. June 1969. Comment on Sternglass thesis. *Bulletin of the Atomic Scientists* 25:27.

Eisenbud, Merrill. January 1968. Radioactivity in the environment—radioiodine concentrations in fetal thyroid. *Pediatrics* (Supplement) 41:174–190, Part II.

Finkel, Miriam P., and Biskis, Biruté O. May 1969. Pathological consequences of radiostrontium administered to fetal and infant dogs. *Radiation biology of the fetal and juvenile mammal,* AEC Symposium Series, vol. 17, Proceedings of the 9th Hanford Biology Symposium, pp. 543–565.

Fliedner, T. M., et al. May 1969. Radiological effects produced in pregnant rats and their offspring by continuous infusion of tritiated thymidime. *Radiation biology of the fetal and juvenile mammal,* AEC Symposium Series, vol. 17, Proceedings of the 9th Hanford Biology Symposium, pp. 263–282.

Gentry, J. T., Parkhurst, E., and Bulin, G. V. 1959. An epidemiological study of congenital malformations in New York State. *American Journal of Public Health* 49:497–513.

Gibson, R. W., Bross, I. D., Graham, S., Lillienfeld, A. M., Schuman, L. M., Levin, M. L., and Dowd, J. E. 1968. Leukemia in children exposed to multiple risk factors. *New England Journal of Medicine* 279:906.

Gofman, John W., and Tamplin, Arthur R. October 1970. The radiation effects controversy. *Bulletin of the Atomic Scientists* 26:2.

Graham, S., Levin, M. L., Lillienfeld, A. M., Schuman, L. M., Gibson,

R. G., Dowd, J. E., and Hempelman, L. 1966. Preconception, intrauterine, and postnatal irradiation as related to leukemia. *National Cancer Institute Monograph* 19:347–371.

Graul, E. H., and Hundeshagen, H. 1958. Studies of the organ distribution of yttrium-90. *Strahlentherapie* 106:405–457.

Griem, M. L., Mehwissen, D. J., Meier, P., and Dobben, G. D. May 1969. Analysis of the morbidity and mortality of children irradiated in fetal life. *Radiation biology of the fetal and juvenile mammal,* AEC Symposium Series, vol. 17, Proceedings of the 9th Hanford Biology Symposium, pp. 651–660.

Harley, John H. 1962. *Report of the United Nations Scientific Commission on the Effects of Radiation.* no. 16A/5216, p. 255, ref. 445.

———. Oct. 1, 1969. Comments on fallout correlations made by Dr. Ernest Sternglass. *Fallout program quarterly summary report,* U.S.A.E.C. Health and Safety Laboratory, New York (HASL-214), vol. I, pp, 2–7.

Hoshino, T., Kato, H., Finch, S., and Hrubec, Z. 1976. Leukemia in offspring of atomic bomb survivors. *Blood* 30:719–730.

Hull, Andrew P. May 15, 1970. Background radiation levels at Brookhaven National Laboratory. Report submitted at the Licensing Hearings, Shoreham Nuclear Plant (AEC Docket No. 50-322).

Illinois Department of Health, Springfield, Ill. 1963–1968. *Illinois vital statistics.* Table D, "Ten leading causes of infant death."

Knelson, J. H. March 1971. Environmental influence on intrauterine lung development. *Archives of Internal Medicine* 127:421–425.

Lade, James H. Nov. 9, 1962. Effect of 1955 fallout in Troy, New York, upon milk and children's thyroids. *Science* 138:732–733.

———. Sept. 13, 1963. That 1953 fallout. *Science* 141:1109–1111.

———. Mar. 6, 1964. More on the 1953 fallout in Troy. *Science* 143:994–995.

Lamont, Lansing. 1965. *Day of Trinity.* New York: Atheneum Publishers.

Lapp, Ralph E. 1955. Global fallout. *Bulletin of the Atomic Scientists* 11:339–343.

———. Sept. 7, 1962. Nevada test fallout and radioiodine in milk. *Science* 137:756–758.

———. Oct. 25, 1963. Nevada test fallout. *Science* 142:448.

Lave, Lester B., Leinhardt, Samuel, and Kaye, Martin B. July 1971. Low-level radiation and U.S. mortality. Working paper no. 19-70-1, Graduate School of Industrial Administration, Carnegie-Mellon University, Pittsburgh.

Le Vann, L. J. 1963. Congenital abnormalities in children born in Alberta during 1961: A survey. *Canadian Medical Association Journal* 89:120–126.

Lewis, E. B. May 17, 1957. Leukemia and ionizing radiation. *Science* 125:965–972.

———. Dec. 13, 1963. Leukemia, multiple myeloma and aplastic anemia in American radiologists. *Science* 142:1492–1494.

Luning, K. G., Frolen, H., Nelson, A., and Ronnback, C. 1963. Genetic effect of strontium-90 injected into male mice. *Nature* (London) 197:304–305.

———. 1963. Genetic effect of strontium-90 on immature germ cells in mice. *Nature* (London) 199:303–304.

MacMahon, B. 1962. Prenatal x-ray exposure and childhood cancers. *Journal of the National Cancer Institute* 28:1173–1191.

Major activities in the atomic energy program. Semiannual Reports, January 1952, and later years.

Martell, E. A. 1974. Radioactivity of tobacco trichomes and insoluble cigarette smoke particles. *Nature* 249:215.

———. 1975. Tobacco radioactivity and cancer in smokers. *American Scientists* 63:404.

May, M. J., and Stuart, I. F. 1970. Comparison of calculated and measured long term gamma doses from a stack effluent of radioactive gases. *Environmental surveillance in the vicinity of nuclear facilities,* p. 234. Springfield, Ill.: Charles C Thomas, Publisher.

Mays, C. W. August 1963. Iodine-131 in Utah during July and August 1962. *Hearings on fallout, radiation standards, and countermeasures.* Joint Committee on Atomic Energy, Part 2, pp. 536–563. Also in Pendleton, R. C., Lloyd, R. D., and Mays, C. W. August 16, 1963. Iodine-131 in Utah during July and August 1962. *Science* 141:640–642.

Moriyama, I. M. May 1960. Recent changes in infant mortality trend. *Public Health Reports* 75(5):391–406.

———. March 1964. The change in mortality trend in the United States. *National Center for Health Statistics,* ser. 3, no. 1, pp. 1–45.

Moskalev, Y. I., et al. May 1969. Experimental study of radionuclide transfer through the placenta and their biological action on the fetus. *Radiation biology of the fetal and juvenile mammal.* AEC Symposium Series, vol. 17, Proceedings of the 9th Hanford Biology Symposium, pp. 153–166.

Müller, W. A. 1967. Gonad dose in male mice after incorporation of strontium-90. *Nature* (London) 214:931–933.

Neel, J. V. 1963. *Changing perspective on the genetic effects of radiation.* Springfield, Ill.: Charles C Thomas, Publisher.

New York State Department of Health, *Annual Statistical Reports.* Hollis S. Ingraham, Commissioner, Albany, New York (available through 1969 and in brief summary form only after 1969).

Petersen, N. J., Samuels, L. C., Lucas, H. F., and Abrahams, S. P. 1966. An epidemiological approach to low-level radium-226 exposure. *Public Health Reports* 81:805–814.

Petkau, A. 1972. Effect of $^{22}Na^+$ on a phospholipid membrane, *Health Physics,* 22:239.

―――― and Chelack, W. S. 1974. Radioprotective effects of cysteine, *International Journal of Radiobiology,* 25:321.

――――, ――――, Pleskach, S. D., and Copps, T. P. 1975. *Radioprotection of hematopoietic and mature blood cells by superoxide dismutase.* Paper presented at the Annual Meeting of the Biophysical Society, Philadelphia. See also: 1976. *Biochimica et Biophysica Acta* 433:445.

Pfeiffer, E. W. September 1965. Mandan milk mystery. *Scientist and Citizen* 7:1–5.

Radiological health data and reports, published monthly by the Bureau of Radiological Health, Rockville, Md. (Published by the Environmental Protection Agency until December 1974.)

Reiss, E. August 1963. *Hearings on fallout radiation standards and countermeasures.* Joint Committee on Atomic Energy, Part 2, pp. 601–672.

Report of the United Nations Scientific Committee on the effects of atomic radiation. Annex G. United Nations, New York, 1962.

Rosenthal, Harold L. May 1969. Accumulation of environmental Sr-90 in teeth of children. *Radiation biology of the fetal and juvenile mammal.* AEC Symposium Series, vol. 17, Proceedings of the 9th Annual Hanford Biology Symposium, pp. 163–171.

Rotblat, J. 1955. The hydrogen-uranium bomb. *Bulletin of the Atomic Scientists* 11:171–177.

Sagan, L. A. May 1969. Human effects of low-level radiation: a critique. *Radiation biology of the fetal and juvenile mammal.* AEC Symposium Series, vol. 17, Proceedings of the 9th Annual Hanford Biology Symposium, pp. 719–725.

――――. October 1969. The infant mortality controversy. *Bulletin of the Atomic Scientists* 25:26–28.

――――. Oct. 2, 1969. A reply to Sternglass. *New Scientist* 44:14–18.

Shleien, B. May 1970. *An estimate of radiation doses received in vicinity of a nuclear fuel reprocessing plant.* U.S. Department of Health, Education and Welfare, Bureau of Radiological Health, Rockville, Md. (BRH-NERHL 70-1).

Spode, E. 1958. On the distribution of radioyttrium and radioactive rare-earth elements in mammals. *Naturforschung* 13b:286–291.

Sternglass, E. J. June 7, 1963. Cancer: relation of prenatal radiation to development of the disease in childhood. *Science* 140:1102–1104.

——. April 1969. Infant mortality and nuclear tests. *Bulletin of the Atomic Scientists* 25:18–20.

——. June 1969. Can the infants survive? *Bulletin of the Atomic Scientists* 25:26.

——. September 1969. The death of all children. *Esquire.*

——. December 1969. Evidence for low-level radiation effects on the human embryo and fetus. *Radiation biology of the fetal and juvenile mammal.* AEC Symposium Series, vol. 17, Proceedings of 9th Hanford Biology Symposium, pp. 693–717.

——. December 1969. A reply. *Bulletin of the Atomic Scientists* 25:29–34.

——. May 1970. A reply. *Bulletin of the Atomic Scientists* 25:41–42, 47.

——. September 1970. Infant mortality and nuclear testing: a reply. *Quarterly Bulletin of the American Association of Physicists in Medicine* 4:115–119.

——. November 1970. Infant mortality changes near a nuclear fuel reprocessing facility. University of Pittsburgh (unpublished).

——. 1972. Environmental radiation and human health. *Proceedings of the 6th Berkeley Symposium on Mathematical Statistics and Probability,* Berkeley, Calif., July 19–22, 1971, ed: J. Neyman. Berkeley: University of California Press.

——. 1974. Nuclear radiation and human health. Proceedings of the International Pollution Control Conference, University of Trondheim, Trondheim, Norway, Aug. 26–29, 1971. Oslo, Norway: University of Oslo Press. Also in *Against pollution and hunger* ed. A. M. Hilton. New York: Halsted Press, 1974.

——. 1976. The role of indirect radiation effects on cell membranes in the immune response. *Radiation and the immune process,* Proceedings of the 1974 Hanford Radiobiology Symposium, Division of Technical Information, ERDA, Oak Ridge, Tenn. (Conf-740930).

Stewart, Alice M. May 1969. Radiogenic cancers of childhood. *Radiation biology of the fetal and juvenile mammal.* AEC Symposium

Series, vol. 17, Proceedings of the 9th Annual Hanford Biology Symposium, pp. 681–692.

―――― and Hewitt, David. 1965. Leukemia incidence in children in relation to radiation exposure in early life. *Current topics in radiation research,* ed. Michael Ebert and Alma Howard, vol. I. Amsterdam: North-Holland Publishing Company.

―――― and Kneale, G. W. June 6, 1970. Radiation dose effects in relation to obstetric x-rays and childhood cancers. *Lancet* 1:1185–1188.

――――, Webb, J., and Hewitt, David. 1958. A survey of childhood malignancies. *British Medical Journal* 1:1495–1508.

Tamplin, Arthur R. December 1969. Infant mortality and the environment. *Bulletin of the Atomic Scientists* 25:23–29.

Tompkins, Edythalena. 1971. Infant mortality around three nuclear reactors. *Proceedings of the 6th Berkeley Symposium on Mathematical Statistics and Probability,* Berkeley, Calif., July 19–22, 1971, ed. J. Neyman. Berkeley: University of California Press.

United Nations Scientific Committee on the Effects of Radiation, 24th Session, Supplement No. 13 (A/7613), 1969.

U.S. Public Health Service. March 1970. *Radiological surveillance studies at a boiling water nuclear power station.* Department of Health, Education and Welfare, Bureau of Radiological Health, Rockville, Md. (BRH-DER 70-1).

――――. March 1970. *Radioactive waste discharges to the environment from nuclear power facilities.* Department of Health, Education and Welfare, Bureau of Radiological Health, Rockville, Md. (BRH-DER 70-2).

Weiss, E. S., Olsen, R. E., Thompson, G. D. C., and Masi, A. T. September 1967. Surgically treated thyroid disease among young people in Utah, 1948–1962. *American Journal of Public Health* 57:1807–1814.

GENERAL REFERENCES UTILIZED

I. U.S. Government Publications and Hearings Relating to Fallout, Radiation Standards, Nuclear Weapons Effects, and the Nuclear Test Ban Treaty

Basic radiation protection criteria. Jan. 15, 1971. Recommendations of the National Council on Radiation Protection and Measurements, NCRP Report No. 39 (available from the NCRP, 4201 Connecticut Ave., N.W., Washington, D.C. 20008).

Glasstone, Samuel, ed. April 1962. *The effects of nuclear weapons.* U.S. Department of Defense and U.S. Atomic Energy Commission.

Slade, David H., ed. July 1968. *Meteorology and atomic energy—1968.* U.S. Atomic Energy Commission, USAEC Division of Technical Information, Oak Ridge, Tenn.

U.S. Atomic Energy Commission. January 1953. *Assuring public safety in continental weapons tests.*

U.S. Atomic Energy Commission. *Major activities in the atomic energy programs.* Semiannual and annual reports.

U.S. Congress. *Hearings of the Joint Committee on Atomic Energy, Special Subcommittee on Radiation* (available from the U.S. Government Printing Office, Washington, D.C.): *(a)* May 1957 hearings on fallout; *(b)* May 1959 hearings on fallout; *(c)* June 1959 hearings on the biological and environmental effects of nuclear war; *(d)* June 1962 hearings on fallout; *(e)* June 1963 hearings on fallout; *(f)* August 1963 hearings on fallout; *(g)* June 1965 hearings on federal radiation council protective action guides.

U.S. Congress. Part 1 (October–November 1969) and Part 2 (January–February 1970). *Environmental effects of producing electric power.* Hearings before the Joint Committee on Atomic Energy, 91st Congress.

U.S. Senate. Aug. 12–27, 1963. *Nuclear test ban treaty.* Hearings before the Committee on Foreign Relations, 88th Congress.

U.S. Senate. Nov. 18–20, 1969. *Underground uses of nuclear energy.* Hearings before the Subcommittee on Air and Water Pollution, Committee on Public Works, 91st Congress.

II. Biological Hazards of Nuclear Radiation

Alexander, P. 1965. *Atomic radiation and life.* London: Penguin Books, Inc.

BEIR Report, National Academy of Sciences, Washington, D.C. November, 1972 and 1979. *The effects on populations of exposure to low levels of ionizing radiation.*

Brodine, Virginia. 1975. *Radioactive contamination.* New York: Harcourt Brace Jovanovich.

Commoner, Barry. 1963. *Science and survival.* New York: The Viking Press, Inc.

Gofman, John W., and Tamplin, Arthur R. 1970. *Population control through nuclear pollution.* Chicago: Nelson-Hall Company.

Lapp, Ralph E. 1958. *The voyage of the Lucky Dragon.* New York: Harper & Brothers.

Lea, D. E. 1962. *Actions of radiations on living cells.* London: Cambridge University Press.

Lundin, F., Wagoner, J. K., and Archer, V. E. 1971. *Radon daughter exposure and respiratory cancer,* Joint Monography # 1, National Institute for Occupational Safety and Health, and National Institute of Environmental Health Sciences. U.S. Department of Health, Education and Welfare (PB-204-871), National Technical Information Service, Springfield, Va., 22151.

Pauling, Linus. 1958. *No more war.* New York: Dodd, Mead & Company, Inc.

Sakharov, A. D. June 1958. Radioactive carbon from nuclear explosions and nonthreshold biological effects. *Soviet Journal of Atomic Energy,* vol. 4, no. 6.

Schubert, Jack, and Lapp, Ralph E. 1958. *Radiation: what it is and how it affects you.* New York: The Viking Press, Inc.

Scientist and Citizen (now *Environment*), St. Louis Committee for Environmental Information, 438 N. Skinker Blvd., St. Louis, Mo. 63130. Series of articles on the hazard of nuclear fallout in the following issues: July 1962; October–November 1962; December 1962; February 1963; May 1963; July 1963; September–October 1963; February 1964; March 1964; June–July 1964; May–June 1965; August 1965; February–March 1966 (back issues available from Maxwell Reprint Co., Fairview Park, Elmsford, N.Y. 10532).

Sikov, Melvin R., and Mahlum, D. Dennis, eds. May 5–8, 1969. *Radiation biology of the fetal and juvenile mammal.* Proceedings of the 9th Hanford Biology Symposium, U.S. Atomic Energy Commission, Division of Technical Information, Oak Ridge, Tenn. (available as CONF-690501, Clearing House for Federal Scientific and Technical Information, Springfield, Va. 22151).

Sternglass, E. J. 1972. Environmental radiation and human health, in *Effects of pollution on health,* vol. 6, in *Proceedings of the 6th*

Berkeley Symposium on Mathematical Statistics and Probability, Berkeley, Calif., ed. L. M. LeCam, J. Neyman, and E. L. Scott. Berkeley: University of California Press, pp. 145–221.

———. June 1973. *Epidemiological studies of fallout and patterns of cancer mortality*, AEC Symposium Series, vol. 29, Proceedings of the Hanford Biology Symposium (Conf-720505).

———. 1977. Radioactivity, chap. XV in *Environmental Chemistry*, ed. J. O. Bockris. New York: Plenum Press.

United Nations. 1958, 1962, 1964, 1966, 1969. *Reports of the United Nations Scientific Committee on the effects of radiation.* The 1969 report deals especially with the evidence for low-dose radiation effects on human chromosomes and the central nervous system.

U.S. Atomic Energy Commission. October 1969. *Inhalation carcinogenesis.* Symposium Series, vol. 18, Proceedings of a Biology Division, Oak Ridge National Laboratory, Conference, USAEC Division of Technical Information, Oak Ridge, Tenn.

U.S. Congress. May–June 1957 (2 vols.) and May 1959 (2 vols.). *The nature of radioactive fallout and its effect on man.* Hearings of the Joint Committee on Atomic Energy, 85th Congress. Washington, D.C.: U.S. Government Printing Office.

III. Trends in Mortality and Statistical Sources

Moriyama, I. M. 1960. Recent change in infant mortality trend. *Public Health Report* 75:391–405.

———. 1964. *The change in mortality trend in the United States.* National Center for Health Statistics, Report No. 1, Ser. 3.

Segi, M., Kurihara, M., and Matsuyama, T. 1965. *Cancer mortality in Japan, 1899–1962.* (Also later supplements.) Department of Public Health, Tohoku University School of Medicine, Sendai, Japan.

Shapiro, S., Schlesinger, E. R., and Nesbitt, R. E. L., Jr. 1968. *Infant, perinatal, maternal, and childhood mortality in the United States.* Cambridge: Harvard University Press.

U.S. Bureau of the Census. 1970 and earlier editions. *Statistical abstract of the United States.* U.S. Department of Commerce, Washington, D.C.

U.S. Public Health Service. November 1965. *Infant mortality trends—United States and each state, 1930–1964.* National Center for Health Statistics, Report No. 1, Ser. 20.

_____. November 1965. *Changes in mortality trends in Chile.* National Center for Health Statistics, Report No. 2, Ser. 3.

_____. November 1965. *Changes in mortality trends in England and Wales, 1931–1961.* National Center for Health Statistics, Report No. 3, Ser. 3.

_____. June 1966. *Report of the international conference on the perinatal and infant mortality problem of the United States.* National Center for Health Statistics, Report No. 3, Ser. 4.

_____. March 1967. *International comparison of perinatal and infant mortality—the United States and six West European countries.* National Center for Health Statistics, Report No. 6, Ser. 3.

_____. June 1968. *Recent retardation of mortality trends in Japan.* National Center for Health Statistics, Report No. 10, Ser. 3.

_____. *Monthly vital statistics reports.* Department of Health, Education and Welfare, National Center for Health Statistics, Rockville, Md. 20852.

_____. *Annual volumes, vital statistics of the United States.* Department of Health, Education and Welfare, National Center for Health Statistics, Rockville, Md. 20852.

World health statistics annual. Vital statistics and causes of death. World Health Organization, Geneva, Switzerland. (Data for fifty-four countries.)

IV. Studies of Low-Level Radiation Effects Produced by X-rays or Natural Background Radiation in Human Populations

Barcinski, M. A. July 15, 1971. Chromosome analysis of population in Brazilian areas of high natural radioactivity. Presented at the Annual Meeting, Health Physics Society, New York. Unpublished Report of the Institute of Biophysics, Rio de Janeiro, Brazil.

Bertel, R. 1979. *Maternal child health effects from radioactive particles in milk, Wisconsin 1963–75,* American Public Health Association Annual Meeting, New York.

Bross, I. D. J., and Natarajian, N. 1972. *Leukemia from low-level radiation, New England Journal of Medicine.* 287:107.

Diamond, E. I., Schmerler, H., and Lillienfeld, A. M. 1973. *The relationship of intrauterine radiation to subsequent mortality and development of leukemia in children, American Journal of Epidemiology,* 97:283.

Gentry, J. T., Parkhurst, E., and Bulin, G. V. 1959. An epidemiological study of congenital malformations in New York State, *American Journal of Public Health,* 49:497.

Graham, S., Levin, M. L., Lillienfeld, A. M., Schuman, L. M., Gibson, R. G., Dowd, J. E., and Hempelman, L. 1966. Preconception intrauterine, and postnatal irradiation as related to leukemia. *National Cancer Institute, Monograph* 19:347–371.

Jacobsen, Lars. 1968. Low dose x-ray radiation and teratogenesis. Copenhagen: Munksgaard Publishing Company.

Kochupilai, N., Verma, I. C., Grewai, M. S., and Ramalinga, Swami, V. July 1, 1976. Down's syndrome and related abnormalities in an area of high background radiation in coastal Kerdla, *Nature,* 262.

Lewis, E. B. May 15, 1957. Leukemia and ionizing radiation. *Science* 125:965–972.

_____. Dec. 13, 1963. Leukemia, multiple myeloma, and aplastic anemia in American radiologists. 142:1492–1494.

Macht, S. H., and Lawrence, P. S. 1955. National survey of congenital malformations resulting from exposure to roentgen radiation. 73:442–466.

MacMahon, Brian. 1962. Prenatal x-ray exposure and childhood cancer. *Journal of National Cancer Institute* 28:1173–1191.

Mancuso, T. F., Stewart, A., and Kneale, G. 1977. Radiation exposures of Hanford workers dying from various causes. *Health Physics* 33:369.

March, H. C. 1944. Leukemia in radiologists. *Radiology* 43:275–278.

_____. 1950. *American Journal of the Medical Sciences* 220:282.

Peterson, N. J., Samuels, L. D., Lucas, H. F., and Abrahams, S. P. September 1966. An epidemiological approach to low-level radium-226 exposure. *Public Health Reports* 81:805–814.

Rugh, Roberts. February 1969. *The effects of ionizing radiation on the developing embryo and fetus.* Seminar Paper No. 007, Bureau of Radiological Health, U.S. Public Health Service, Rockville, Md.

Seltser, Raymond. 1964. Studies of mortality among American radiologists. *Journal of the American Medical Association.* 190:90.

Stewart, A., Pennibacker, W., and Barber, R. Oct. 6, 1962. Adult leukemias and diagnostic x-rays. *British Medical Journal* 2:882–890.

_____, Webb, J., and Hewitt, D. 1958. A survey of childhood malignancies. *British Medical Journal* 1:1495–1508.

_____, _____, _____ and Giles, D. 1956. Studies of children irradiated in utero. *Lancet* 271:447.

Uchida, Irene A., Holunga, Roberta, and Lawler, Carolyn. Nov. 16, 1968. Maternal radiation and chromosomal aberrations. *Lancet.*

United Nations. 1969. *Radiation-induced chromosome aberrations in human cells.* Report of the United Nations Scientific Committee on the Effects of Radiation, Annex C, p. 156 (Supplement No. 13 (A/7613)).

V. Data on Radioactivity in the Environment and Releases from Nuclear Reactors and Fuel Reprocessing Facilities (U.S. Government reports available from the Clearing House for Federal Scientific and Technical Information, Springfield, Va. 22151)

Cochran, J. A., et al. July 1970. *An investigation of airborne radioactive effluent from an operating nuclear fuel reprocessing plant.* U.S. Public Health Service, Bureau of Radiological Health (BRH-NERHL-70-1).

Magno, P., et al. November 1970. *Liquid waste effluents from a nuclear fuel reprocessing plant.* U.S. Public Health Service, Bureau of Radiological Health (BRH-NERHL-70-2).

Radiological health data and reports. 1957–1974. Published monthly by the Environmental Protection Agency (formerly the Bureau of Radiological Health, U.S.P.H.S.), 5600 Fisher's Lane, Rockville, Md. 20852. Contains monthly reports on radioactivity levels in the milk, water, air, and precipitation together with articles relating to environmental radiation.

Reinig, William C., ed. 1970. *Environmental surveillance in the vicinity of nuclear facilities.* Springfield, Ill.: Charles C Thomas, Publisher.

Shleien, B. May 1970. *An estimate of radiation doses received in vicinity of a nuclear fuel reprocessing plant.* U.S. Department of Health, Education and Welfare, Bureau of Radiological Health, Rockville, Md. (BRH-NERHL 70-1); also July 1970, BRH-NERHL-70-3.

U.S. Public Health Service. March 1970. *Radioactive waste discharges to the environment from nuclear power facilities.* Department of Health, Education and Welfare, Bureau of Radiological Health, Rockville, Md. (BRH-DER 70-2).

————. March 1970. Radiological surveillance studies at a boiling water nuclear power station. Department of Health, Education and Welfare, Bureau of Radiological Health, Rockville, Md. (BRH-DER 70-1).

VI. General References on the Hazards of Nuclear Power Generation and its Alternatives

Abrahamson, Dean E. 1970. *Environmental cost of electric power.* Scientist's Institute for Public Information, 30 East 68 Street, New York, N.Y. 10021.

Bryerton, Gene. 1970. *Nuclear dilemma.* New York: Ballantine Books, Inc.

Bulletin of the Atomic Scientists. The energy crisis. Part I, September 1971; Part II, October 1971; Part III, November 1971. The alternatives to fission power are presented in Part II.

Curtis, Richard C., and Hogan, Elizabeth. 1969, *Perils of the peaceful atom.* New York: Ballantine Books, Inc.

Fabricant, Neil, and Hallman, Robert M. 1971. *Toward a rational power policy—energy, politics and pollution.* New York: George Braziller, Inc.

Gofman, John W., and Tamplin, Arthur R. 1971. *Poisoned power—the case against nuclear power plants.* Emmaus, Pennsylvania: Rodale Press.

Honicker vs. Hendrie—1978. *A law suit to end atomic power.* Summertown, Tenn.: The Book Publishing Company.

Inglis, David R. 1973. *Nuclear energy: its physics and its social challenge.* Reading, Mass.: Addison-Wesley Publishing Company.

Keisling, Bill. 1980. *Three Mile Island—turning point.* Seattle: Veritas Books, Inc.

Lovins, Amory B. 1977. Soft energy paths: towards a durable peace. New York: Penguin Books.

Novick, Sheldon. 1969. *The careless atom.* Boston: Houghton Mifflin Company.

Olsen, McKinley C. 1976. *Unacceptable Risk.* New York: Bantam Books, Inc.

Radiation Standards and Public Health 1978, Environmental Policy Institute, 317 Pennsylvania Ave., S.E., Washington, D.C. 20003.

Reader, Mark, et al. 1980. *Atom's eve: ending the nuclear age, an anthology.* New York: McGraw-Hill Book Company.

Rosenberg, Howard L. 1980. *Atomic soldiers.* Boston: Beacon Press.

Schrader-Frechette, K. S. 1980. *Nuclear power and public policy—the social and ethical problems of fission technology.* Boston: Reidel Publishing Company.

Shut down—1979. Nuclear power on trial. Summertown, Tenn.: The Book Publishing Company.

ADDENDA TO BIBLIOGRAPHY

Caldwell, G. G., Kelly, D. B., Heath, C. W., Jr.. October 1980. "Leukemia among participants in military maneuvers at a nuclear bomb test." *Journal of the American Medical Association* 244:1575–1578.

Field, R. W., Field, E. H., Zegers, D. A., and Steucek, G. L. 1980. "Iodine-131 in thyroids of the meadow vole in the vicinity of the Three Mile Island Nuclear Generating Plant." Department of Biology, Millersville State College, Millersville, Pa. 17551 (Submitted to Health Physics).

Franke, B., Krüger, E., Steinhilber-Schwaab, B., Van de Saud, H., and Teufel, D. 1979. "Radiation exposure to the public from radioactive emissions of nuclear power stations." Critical analysis of the official regulatory guides, Institute for Energy and Environmental Research, 6900 Heidelberg, West Germany.

Gong, J. K., Glomski, C. A., and Bruce, A. K. 1980. "Effects of radiation in the I.O.R. range," in Proceedings of the Symposium on Biological Effects, Imaging Techniques and Dosimetry of Ionizing Radiations, Rockville, Md., June 6–8, 1979, U.S. Department of Health and Human Services, HHS Publication No. (FDA) 80–8126.

Gyorgy, Anna. 1979. "No Nukes—Everyone's guide to nuclear power." South End Press, Boston, Mass. 02123.

Köteles, G. J. 1979. "New aspects of cell membrane radiobiology and their impact on radiation protection." *Atomic Energy Review* 17: No. 1 3–30, International Atomic Energy Agency, Vienna, Austria.

Lyon, J. L., Klauber, M. R., Gardner, J. W., et al. 1979. "Childhood leukemia associated with fallout from nuclear testing." *New England Journal of Medicine* 300:317–402.

Modan, B., Mart, H., Baidatz, D., et al. 1974. "Radiation induced head and neck tumors." *Lancet* 1:277–279.

Najarian, W. A., and Colton, T. 1979. "Mortality from leukemia and cancer in shipyard nuclear workers." *Lancet* 1:1018–1020.

――――, Colton, C., Greenberg, E. R., Barron, J. June 1980. "An analysis of deaths among nuclear shipyard workers." Final Report, Contract 80–1574, National Institute for Occupational Safety and Health, U.S. Public Health Service (Submitted for publication in the *New England Journal of Medicine*).

Petkau, A., and Chelack, W. S. 1976. "Radioprotective effect of superoxide dismutase on model phospholipid membranes." *Biochimica and Biophysica Acta* 433:445–456.

Silverman, C. July 1980. "Mental function following scalp irradiation for tinea capitatis in childhood." Symposium on biological effects, imaging techniques and dosimetry of ionizing radiation," p. 36. Rockville, Md., June 6–8, 1979, U.S. Department of Health and Human Services, HHS Publication No. (FDA) 80–8126. U.S. Government Printing Office, Washington, D.C. 20402.

Sternglass, E. J. 1973. Reports to Governor Milton Shapp of Pennsylvania on abnormal radiation levels and mortality rates near the Shippingport Nuclear Power Station, January and May 1973 (Unpublished).

————. 1977. "Radioactivity." Chapter XV in *Environmental Chemistry*. Ed. J. O. Bockris, New York, N.Y., Plenum Press.

————. 1978. "Cancer mortality changes around nuclear facilities in Connecticut." Proceedings of Second Congressional Seminar on Low-Level Ionizing Radiation, February 1978. Published by the Environmental Policy Institute, 317 Pennsylvania Ave., Washington, D.C. 20003.

————, and Bell, S. 1979. "Fallout and the decline of aptitude scores." Presented at the September 1979 Annual Meeting of the American Psychological Association (Submitted for publication in the *Journal of the American Psychological Association*).

————. 1979. "Infant Mortality Changes Following the Three Mile Island Accident." 5th World Congress of Engineers and Architects, Tel Aviv, Israel, December 1979.

Stewart, Alice M., Kueale, G. W., and Mancuso, T. F. 1980. "A cohort study of the cancer risks from radiation to all workers at Hanford." *British Journal of Industrial Medicine* (in publication).

"Time Bomb—A nuclear reader from the Progressive." 1980. Ed. by James Rowen, The Progressive Foundation, Madison, Wis. 53703.

"Announced United States Nuclear Test Statistics." 1976. U.S. Energy Research and Development Administration, Washington, D.C.

U.S. Congress. Office of Technology Assessment. 1979. "The Effects of Nuclear War." U.S. Government Printing Office, Washington, D.C. 20402.

U.S. Congress. "Effect of radiation on human health." Hearings before the Select Committee on Health and the Environment of the Committee on Interstate and Foreign Commerce, House of Representatives, 96th Congress, 2nd Session, Vol. 1, January and February

1978, Serial #95–179 and Vol. 2, July 1978, Serial #95–180. (U.S. Government Printing Office, Washington, D.C. 20402, 1979)

U.S. Department of Health and Human Services, Public Health Service, Bureau of Radiological Health, Rockville, Md. 20857: Proceedings of Symposium on Biological Effects, Imaging Techniques and Dosimetry of Ionizing Radiation, July 1980. HHS Publication No. (FDA) 80–8126.

U.S. Nuclear Regulatory Commission. "Reports on releases of radioactivity in effluents and solid wastes from nuclear power plants," issued for all years since 1972. NUREG–75–001 for 1973 (issued January 1975); NUREG–0077 for 1974 (issued June 1976); NUREG–0218 for 1975 (issued March 1977); NUREG–0367 for 1976 (issued March 1978); T. R. Decker, 12105 MNBB, Washington, D.C. 20555.

Glossary

Alpha rays: Comparatively large, slow particles emitted from the nucleus of an atom. They are easily stopped but can cause great damage if the chemicals emitting them are inhaled or ingested.

Background radiation: Radiation (at a typical rate of 60–100 mrem per year) coming from space or from the earth. It can be both natural and man-made.

Beta rays: Charged particles (electrons) emitted from the nucleus of an atom that are smaller and faster than alpha rays and can penetrate several layers of tissue up to a few millimeters, or a few meters of air.

Curie: A measure of the amount of radiation emitted per second by radioactive chemicals, named after Marie Curie, the discoverer of radium. It is the number of disintegrations taking place each second in 1 gram of radium, leading to the emission of some 37 billion gamma rays or other particles every second.

millicurie—one one-thousandth of a curie.

microcurie—one millionth of a curie.

picocurie—one trillionth of a curie or one micro-microcurie.

Fuel cycle: The sequence of steps needed for the production and combustion of fuel to produce nuclear energy including mining, milling, conversion, enrichment, transportation, and waste storage.

Gamma rays: A very high energy form of radiation similar to X-rays emitted from the nucleus of an atom that can penetrate steel and concrete.

Kiloton: A measure of the power of an atomic bomb, equal to the detonation of 1000 tons of TNT. The first A-bombs had a size of 10–20 kilotons, now regarded as small, tactical weapons.

Megaton: A million tons of TNT in explosive force, or the energy of 1000 kilotons, some 100 times that released by the first atomic bombs.

Rad and millirad: A radiation measure that refers to the energy absorbed per gram of tissue which is equal to about 83% of the Roentgen value. A millirad or mrad is a thousandth of a rad.

Rem and millirem: A radiation measure that reflects the difference in biological damage of the radiation dose produced by different particles. The relation between rad and rem depends on the kind of particle emitting the radiation: for gamma rays, 1 rad = 1 rem; for beta, 1 rad = 10 rem; for alpha, 1 rad = 30 rem.

Roentgen: The original term used for measuring the amount of ionizing radiation incident on the body. It is equal to the quantity of radiation that will produce one electrostatic unit of electricity in one cubic centimeter of dry air at 0° C.

Index